T0296075

Representations of Lie Algebras

This bold and refreshing approach to Lie algebras assumes only modest prerequisites (linear algebra up to the Jordan canonical form and a basic familiarity with groups and rings), yet it reaches a major result in representation theory: the highest-weight classification of irreducible modules of the general linear Lie algebra. The author's exposition is focused on this goal rather than on aiming at the widest generality, and emphasis is placed on explicit calculations with bases and matrices. The book begins with a motivating chapter explaining the context and relevance of Lie algebras and their representations and concludes with a guide to further reading. Numerous examples and exercises with full solutions are included.

Based on the author's own introductory course on Lie algebras, this book has been thoroughly road-tested by advanced undergraduate and beginning graduate students and is also suited to individual readers wanting an introduction to this important area of mathematics.

AUSTRALIAN MATHEMATICAL SOCIETY LECTURE SERIES

1 Introduction to Linear and Convex Programming, N. CAMERON
2 Manifolds and Mechanics, A. JONES, A. GRAY & R. HUTTON
3 Introduction to the Analysis of Metric Spaces, J. R. GILES
4 An Introduction to Mathematical Physiology and Biology, J. MAZUMDAR
5 2-Knots and their Groups, J. HILLMAN
6 The Mathematics of Projectiles in Sport, N. DE MESTRE
7 The Petersen Graph, D. A. HOLTON & J. SHEEHAN
8 Low Rank Representations and Graphs for Sporadic Groups, C. E. PRAEGER & L. H. SOICHER
9 Algebraic Groups and Lie Groups, G. I. LEHRER (ed.)
10 Modelling with Differential and Difference Equations, G. FULFORD, P. FORRESTER & A. JONES
11 Geometric Analysis and Lie Theory in Mathematics and Physics, A. L. CAREY & M. K. MURRAY (eds.)
12 Foundations of Convex Geometry, W. A. COPPEL
13 Introduction to the Analysis of Normed Linear Spaces, J. R. GILES
14 Integral: An Easy Approach after Kurzweil and Henstock, L. P. YEE & R. VYBORNY
15 Geometric Approaches to Differential Equations, P. J. VASSILIOU & I. G. LISLE (eds.)
16 Industrial Mathematics, G. R. FULFORD & P. BROADBRIDGE
17 A Course in Modern Analysis and its Applications, G. COHEN
18 Chaos: A Mathematical Introduction, J. BANKS, V. DRAGAN & A. JONES
19 Quantum Groups, R. STREET
20 Unitary Reflection Groups, G. I. LEHRER & D. E. TAYLOR
21 Lectures on Real Analysis, F. LÁRUSSON

Australian Mathematical Society Lecture Series: 22

Representations of Lie Algebras

An Introduction Through $\mathfrak{gl}_n$

ANTHONY HENDERSON
School of Mathematics and Statistics
University of Sydney

CAMBRIDGE
UNIVERSITY PRESS

University Printing House, Cambridge CB2 8BS, United Kingdom

One Liberty Plaza, 20th Floor, New York, NY 10006, USA

477 Williamstown Road, Port Melbourne, VIC 3207, Australia

314-321, 3rd Floor, Plot 3, Splendor Forum, Jasola District Centre, New Delhi - 110025, India

103 Penang Road, #05-06/07, Visioncrest Commercial, Singapore 238467

Cambridge University Press is part of the University of Cambridge.

It furthers the University's mission by disseminating knowledge in the pursuit of education, learning and research at the highest international levels of excellence.

www.cambridge.org
Information on this title: www.cambridge.org/9781107653610

First published 2012

A catalogue record for this publication is available from the British Library

Library of Congress Cataloging in Publication data
Henderson, Anthony, 1976- author.
Representations of Lie algebras : an introduction through gln / Anthony Henderson, School of Mathematics and Statistics, University of Sydney.
pages cm. – (Australian Mathematical Society lecture series ; 22)
ISBN 978-1-107-65361-0 (pbk.)
1. Representations of Lie algebras. I. Title.
QA252.3.H46 2012
512′.482–dc23
2012021841

ISBN 978-1-107-65361-0 Paperback

Contents

Preface *page* vii
Notational conventions ix

1 Motivation: representations of Lie groups 1
1.1 Homomorphisms of general linear groups 1
1.2 Multilinear algebra 3
1.3 Linearization of the problem 7
1.4 Lie's theorem 10

2 Definition of a Lie algebra 13
2.1 Definition and first examples 13
2.2 Classification and isomorphisms 15
2.3 Exercises 18

3 Basic structure of a Lie algebra 21
3.1 Lie subalgebras 21
3.2 Ideals 24
3.3 Quotients and simple Lie algebras 27
3.4 Exercises 31

4 Modules over a Lie algebra 33
4.1 Definition of a module 33
4.2 Isomorphism of modules 38
4.3 Submodules and irreducible modules 41
4.4 Complete reducibility 44
4.5 Exercises 47

5 The theory of $\mathfrak{sl}_2$-modules 50
5.1 Classification of irreducibles 50
5.2 Complete reducibility 55
5.3 Exercises 57

6 General theory of modules 60
6.1 Duals and tensor products 60
6.2 Hom-spaces and bilinear forms 65
6.3 Schur's lemma and the Killing form 68
6.4 Casimir operators 72
6.5 Exercises 75

7 Integral $\mathfrak{gl}_n$-modules 78
7.1 Integral weights 78
7.2 Highest-weight modules 86
7.3 Irreducibility of highest-weight modules 90
7.4 Tensor-product construction of irreducibles 93
7.5 Complete reducibility 99
7.6 Exercises 104

8 Guide to further reading 106
8.1 Classification of simple Lie algebras 106
8.2 Representations of simple Lie algebras 109
8.3 Characters and bases of representations 111

Appendix Solutions to the exercises 115
Solutions for Chapter 2 exercises 115
Solutions for Chapter 3 exercises 122
Solutions for Chapter 4 exercises 127
Solutions for Chapter 5 exercises 132
Solutions for Chapter 6 exercises 137
Solutions for Chapter 7 exercises 144

References 153
Index 154

Preface

The aim of this book

Why another introduction to Lie algebras? The subject of this book is one of the areas of algebra that has been most written about. The basic theory was unearthed more than a century ago and has been polished in a long chain of textbooks to a sheen of classical perfection. Experts' shelves are graced by the three volumes of Bourbaki [1]; for students with the right background and motivation to learn from them, the expositions in the books by Humphreys [10], Fulton and Harris [6], and Carter [2] could hardly be bettered; and there is a recent undergraduate-level introduction by Erdmann and Wildon [4]. So where is the need for this book?

The answer comes from my own experience in teaching courses on Lie algebras to Australian honours-level undergraduates (see the Acknowledgements section). Such courses typically consist of 24 one-hour lectures. At my own university the algebraic background knowledge of the students would be: linear algebra up to the Jordan canonical form, the basic theory of groups and rings, the rudiments of group representation theory, and a little multilinear algebra in the context of differential forms. From that starting point, I have found it difficult to reach any peak of the theory by following the conventional route. My definition of a peak includes the classification of simple Lie algebras, the highest-weight classification of their modules, and the combinatorics of characters, tensor products, and crystal bases; by 'the conventional route' I mean the path signposted by the theorems of Engel and Lie (about solvability), Cartan (about the Killing form), Weyl (about complete reducibility), and Serre, as in the book by Humphreys [10]. Following that path without skipping proofs always seemed to require more than 24 lectures.

The solution adopted in this book is drastic. I have abandoned the wider class of simple Lie algebras, focusing instead on the general linear Lie algebra $\mathfrak{gl}_n$, which is almost, but not quite, simple. I have jettisoned all five of the aforementioned theorems, in favour of arguments specific to $\mathfrak{gl}_n$, especially the use of explicit Casimir operators. Although these omissions may shock the experts, I have found this to be an approach that is more accessible and yet still reaches one peak: the classification of $\mathfrak{gl}_n$-modules by their highest weights.

I have started the journey with a motivatory chapter, which gives some explanation of why algebraists care about this classification and also introduces some necessary multilinear algebra. Chapters 2 to 4 cover the basic definitions of Lie algebras, homomorphisms and isomorphisms, subalgebras, ideals, quotients, modules, irreducibility and complete reducibility. In a lecture course, the material in these first four chapters would typically take about 12 hours; so the elegant $\mathfrak{sl}_2$ theory in Chapter 5 is reached relatively early. Then in Chapter 6 I return to the theory of modules, covering tensor products, bilinear forms, Schur's lemma, and Casimir operators.

In Chapter 7 these tools are used to develop the highest-weight theory. My hope is that students who reach the end of Chapter 7 will be inspired to progress to more comprehensive books, and Chapter 8 is intended as a map of what lies ahead.

Acknowledgements

This book began life as a set of lecture notes for my Lie algebras course in the 2004 Australian Mathematical Sciences Institute (AMSI) Summer School. It was extensively revised over the next seven years, as I taught the subject again for the summer school and as an honours course at the University of Sydney. Most of the exercises were originally assignment or exam questions.

I would like to thank AMSI for the initial opportunity to teach this beautiful subject, and the students in all those classes for their feedback. I would also like to thank Pramod Achar, Wai Ling Yee, Cheryl Praeger, and the anonymous reviewers for their valuable suggestions and encouraging comments.

Notational conventions

To simplify matters, we make a standing convention:

All vector spaces are over $\mathbb{C}$ and finite-dimensional.

The finite-dimensionality assumption allows the explicit calculations with bases and matrices that are a feature of the book. The $n \times n$ identity matrix is written 1_n, and the identity transformation of a vector space V is written 1_V. The elements of the vector space $\mathbb{C}^n$ are always thought of as column vectors; linear transformations of this particular vector space are tacitly identified with $n \times n$ matrices (multiplying on the left of the vectors). The bases of vector spaces are considered to be ordered sets and hence are written without set braces. The span of the elements $v_1, \dots, v_k$ is written $\mathbb{C}\{v_1, \dots, v_k\}$. The term 'subspace' always means 'sub-vector-space'. If W and W' are subspaces of a larger vector space V then $W \oplus W'$ denotes their sum, and it is implied that $W \cap W' = \{0\}$ (an 'internal direct sum'); if W and W' are not subspaces of a larger vector space V then $W \oplus W'$ means the 'external direct sum' $\{(w, w') \mid w \in W, w' \in W'\}$. The same principles apply to direct sums with more than two summands.

On its rare appearances, the square root of -1 is written $\mathbf{i}$ to distinguish it from the italic letter i, which is widely used for other purposes. The group of nonzero complex numbers is written $\mathbb{C}^\times$. The set of nonnegative integers is written $\mathbb{N}$. Other notation will be explained as it is needed.

CHAPTER 1

Motivation: representations of Lie groups

Sophus Lie was a Norwegian mathematician who lived from 1842 to 1899. Essentially single-handedly he discovered two fundamental classes of objects in modern mathematics, which now bear his name: Lie groups and Lie algebras. More importantly, he built a bridge between them; this is remarkable, because Lie groups seem to be part of differential geometry (in today's language) while Lie algebras seem to be purely algebraic. In this chapter we will discuss a small part of Lie's discovery.

1.1 Homomorphisms of general linear groups

Typically, Lie groups are infinite groups whose elements are invertible matrices with real or complex entries. So they are subgroups of the *general linear group*

$$GL_n = \{g \in \mathrm{Mat}_n \mid \det(g) \neq 0\},$$

where $\mathrm{Mat}_n = \mathrm{Mat}_n(\mathbb{C})$ denotes the set of $n \times n$ complex matrices for some positive integer n. Lie was interested in such groups because they give the symmetries of differential equations, but they have since found many other applications in areas such as differential geometry and harmonic analysis.

One of the most important algebraic problems concerning Lie groups is to classify a suitable class of *matrix representations* of a given Lie group G, i.e. group homomorphisms $G \to GL_m$ for various m. For the purposes of motivation, we concentrate on the case where G is the full general linear group GL_n; thus the problem can be stated (vaguely) as follows.

Problem 1.1.1. Describe all group homomorphisms $\Phi : GL_n \to GL_m$.

By definition, such a homomorphism is a map $\Phi : GL_n \to \mathrm{Mat}_m$ such that:

$$\Phi(1_n) = 1_m, \tag{1.1.1}$$

where 1_n denotes the $n \times n$ identity matrix, and

$$\Phi(gh) = \Phi(g)\Phi(h) \quad \text{for all } g, h \in GL_n. \tag{1.1.2}$$

(The case $h = g^{-1}$ of (1.1.2), combined with (1.1.1), forces $\Phi(g)$ to be invertible.) Such a map Φ is a collection of m^2 functions $\Phi_{ij} : GL_n \to \mathbb{C}$, where $\Phi_{ij}(g)$ is the (i, j) entry of the matrix $\Phi(g)$. Each function Φ_{ij} is in effect a function of n^2 variables, the entries of the input matrix g (the given domain consists of just the invertible matrices, so the function may or may not be defined for those choices of variables that give a zero determinant). So (1.1.1) and (1.1.2) amount to a complicated system of functional equations. To frame Problem 1.1.1 rigorously, we would have to specify what kinds of function are allowed as solutions – for example, continuous, differentiable, rational, or polynomial – but we will leave this undetermined for now and see what happens in some examples.

Example 1.1.2. The determinant $\det : GL_n \to \mathbb{C}^\times$ is one such homomorphism, if we make the obvious identification of $\mathbb{C}^\times$ with GL_1. The determinant of a matrix is clearly a polynomial function of the entries. ∎

Example 1.1.3. The transpose map $GL_n \to GL_n : g \mapsto g^{\mathrm{t}}$ is not an example because it is an anti-automorphism rather than an automorphism: $(gh)^{\mathrm{t}}$ equals $h^{\mathrm{t}} g^{\mathrm{t}}$ and doesn't usually equal $g^{\mathrm{t}} h^{\mathrm{t}}$. But this means that the map $GL_n \to GL_n : g \mapsto (g^{\mathrm{t}})^{-1}$ is an example. The entries of $(g^{\mathrm{t}})^{-1}$ are rational functions of the entries of g: they are quotients of various $(n-1) \times (n-1)$ minors and the determinant. Therefore these functions are not defined on matrices with zero determinant. ∎

Example 1.1.4. A map that is easily seen to satisfy (1.1.1) and (1.1.2) is the 'duplication' map

$$GL_2 \to GL_4 : \begin{pmatrix} a & b \\ c & d \end{pmatrix} \mapsto \begin{pmatrix} a & b & 0 & 0 \\ c & d & 0 & 0 \\ 0 & 0 & a & b \\ 0 & 0 & c & d \end{pmatrix}.$$

To produce something less trivial-looking, we could replace either copy of $\left(\begin{smallmatrix} a & b \\ c & d \end{smallmatrix}\right)$ with its conjugate $X\left(\begin{smallmatrix} a & b \\ c & d \end{smallmatrix}\right)X^{-1}$, for some fixed $X \in GL_2$, or indeed we could conjugate the whole output matrix by some fixed $Y \in GL_4$. This is a superficial change that we could account for by introducing a suitable equivalence relation into the statement of Problem 1.1.1. Note that in this example the entries of the output matrix are linear functions of the entries of the input matrix. ∎

Example 1.1.5. More interesting is the map $\Psi : GL_2 \to GL_3$ defined by

$$\Psi \begin{pmatrix} a & b \\ c & d \end{pmatrix} = \begin{pmatrix} a^2 & 2ab & b^2 \\ ac & ad + bc & bd \\ c^2 & 2cd & d^2 \end{pmatrix},$$

where the entries of the output are homogeneous polynomials of degree 2 in the entries of the input. It is clear that property (1.1.1) is satisfied. The proof of property (1.1.2) is as follows:

$$\Psi\begin{pmatrix} a & b \\ c & d \end{pmatrix}\Psi\begin{pmatrix} e & f \\ g & h \end{pmatrix} = \begin{pmatrix} a^2 & 2ab & b^2 \\ ac & ad+bc & bd \\ c^2 & 2cd & d^2 \end{pmatrix}\begin{pmatrix} e^2 & 2ef & f^2 \\ eg & eh+fg & fh \\ g^2 & 2gh & h^2 \end{pmatrix}$$

$$= \begin{pmatrix} (ae+bg)^2 & 2(ae+bg)(af+bh) & (af+bh)^2 \\ (ae+bg)(ce+dg) & \begin{matrix} (ae+bg)(cf+dh) \\ +(af+bh)(ce+dg) \end{matrix} & (af+bh)(cf+dh) \\ (ce+dg)^2 & 2(ce+dg)(cf+dh) & (cf+dh)^2 \end{pmatrix}$$

$$= \Psi\left(\begin{pmatrix} a & b \\ c & d \end{pmatrix}\begin{pmatrix} e & f \\ g & h \end{pmatrix}\right).$$

At the moment this seems like an accident, and it is not clear how to find other such solutions of (1.1.1) and (1.1.2). ■

1.2 Multilinear algebra

The right context for explaining the above examples of homomorphisms, and for finding new examples, is the theory of multilinear algebra. If V is an n-dimensional vector space with chosen basis $v_1, \ldots, v_n$ then the elements of GL_n correspond bijectively to invertible linear transformations of V: a matrix (a_{ij}) in GL_n corresponds to the unique linear map $\tau : V \to V$ such that

$$\tau(v_j) = \sum_{i=1}^{n} a_{ij} v_i \quad \text{for all } j. \tag{1.2.1}$$

If we have a way of constructing from V a new vector space W with basis $w_1, \ldots, w_m$, and if this construction is sufficiently 'natural', then each linear transformation of V should induce a linear transformation of W and the resulting map $\Phi : GL_n \to GL_m$ of matrices should satisfy (1.1.1) and (1.1.2). This is one reason to be interested in Problem 1.1.1: the homomorphisms between general linear groups tell us something about natural constructions of vector spaces.

Example 1.2.1. A very important example of such a homomorphism occurs when W is the *dual space* V^*, consisting of all linear functions $f : V \to \mathbb{C}$. This is also n-dimensional: it has a basis $v_1^*, \ldots, v_n^*$, where v_i^* is the unique linear function satisfying

$$v_i^*(v_j) = \delta_{ij} = \begin{cases} 1 & \text{if } i = j, \\ 0 & \text{otherwise.} \end{cases} \tag{1.2.2}$$

In other words, v_i^* is the function whose value on $a_1v_1 + \cdots + a_nv_n \in V$ is the coefficient a_i. A general linear function $f : V \to \mathbb{C}$ can be written as $f(v_1)v_1^* + \cdots + f(v_n)v_n^*$. If τ is an invertible linear transformation of V then τ induces in a natural way an invertible linear transformation τ^* of V^*, defined by

$$\tau^*(f)(v) = f(\tau^{-1}(v)) \quad \text{for all } v \in V, f \in V^*. \tag{1.2.3}$$

(The transformation τ^{-1} on the right-hand side does indeed give the function that one would naturally expect, for the same reason that, in calculus, translating the graph of $y = f(x)$ one unit to the right gives the graph of $y = f(x-1)$.) To find the matrix of τ^* relative to the basis $v_1^*, \ldots, v_n^*$, observe that its (j, i) entry is the coefficient of v_j^* in $\tau^*(v_i^*)$; this is the same as $\tau^*(v_i^*)(v_j) = v_i^*(\tau^{-1}(v_j))$, the coefficient of v_i in $\tau^{-1}(v_j)$, i.e. the (i, j) entry of the matrix of τ^{-1} relative to $v_1, \ldots, v_n$. So, the map of matrices corresponding to $\tau \mapsto \tau^*$ is the inverse transpose map considered in Example 1.1.3. ■

Example 1.2.2. Take $W = V \oplus V = \{(v, v') \mid v, v' \in V\}$. Any linear transformation τ of V induces a linear transformation $\tau \oplus \tau$ of $V \oplus V$, defined by

$$(\tau \oplus \tau)(v, v') = (\tau(v), \tau(v')) \quad \text{for all } v, v' \in V. \tag{1.2.4}$$

The most obvious basis for $V \oplus V$ consists of

$$(v_1, 0), (v_2, 0), \ldots, (v_n, 0), (0, v_1), (0, v_2), \ldots, (0, v_n).$$

Relative to this basis, the matrix corresponding to $\tau \oplus \tau$ is exactly the block-diagonal duplication of the matrix of τ seen in Example 1.1.4; the conjugated versions mentioned there would arise if one used other bases of $V \oplus V$. ■

To explain Examples 1.1.2 and 1.1.5 similarly, we need the concept of the *tensor product*, which for finite-dimensional vector spaces can be explained fairly simply. Given two vector spaces V and W with respective bases $v_1, \ldots, v_n$ and $w_1, \ldots, w_m$, the tensor product $V \otimes W$ is a vector space with basis $v_i \otimes w_j$ for all i, j with $1 \leq i \leq n$, $1 \leq j \leq m$. One can regard the elements $v_i \otimes w_j$ merely as symbols and $V \otimes W$ as the space of formal linear combinations of them. Note that the dimension of $V \otimes W$ is $(\dim V)(\dim W)$, in contrast with that of the direct sum $V \oplus W$, which is $\dim V + \dim W$. For arbitrary elements $v \in V$ and $w \in W$, we define the *pure tensor* $v \otimes w \in V \otimes W$ by the following rule:

$$\begin{aligned} &\text{if } v = a_1v_1 + \cdots + a_nv_n \quad \text{and} \quad w = b_1w_1 + \cdots + b_mw_m \\ &\text{then } v \otimes w = \sum_{i=1}^{n} \sum_{j=1}^{m} a_ib_j(v_i \otimes w_j). \end{aligned} \tag{1.2.5}$$

Note that if v happens to equal v_i and w happens to equal w_j then $v \otimes w$ does indeed equal the basis element $v_i \otimes w_j$, so our notation is consistent. Having made this definition, one can easily show that the tensor product does not depend on the chosen bases of V and W: for any other bases $v_1', \ldots, v_n'$ and $w_1', \ldots, w_m'$ the elements $v_i' \otimes w_j'$ form another, equally good, basis of $V \otimes W$. It is important to bear in mind that a general element of $V \otimes W$ is not a pure tensor: it is, of course, a linear combination of the basis elements $v_i \otimes w_j$ but the coefficients cannot usually be written in the form $a_i b_j$, as in (1.2.5).

So, we have another way to construct a new vector space from a vector space V: we can consider its tensor square $V^{\otimes 2} = V \otimes V$. Any linear transformation τ of V induces a linear transformation $\tau \otimes \tau$ of $V \otimes V$, defined on the basis elements by $(\tau \otimes \tau)(v_i \otimes v_j) = \tau(v_i) \otimes \tau(v_j)$. It is easy to see that in fact

$$(\tau \otimes \tau)(v \otimes v') = \tau(v) \otimes \tau(v') \quad \text{for any } v, v' \in V. \tag{1.2.6}$$

Example 1.2.3. Suppose that V is two-dimensional, with basis v_1, v_2. If the linear transformation $\tau : V \to V$ has matrix $\left(\begin{smallmatrix} a & b \\ c & d \end{smallmatrix}\right)$ relative to this basis then, for instance,

$$\begin{aligned}(\tau \otimes \tau)(v_1 \otimes v_1) &= \tau(v_1) \otimes \tau(v_1) \\ &= (av_1 + cv_2) \otimes (av_1 + cv_2) \\ &= a^2(v_1 \otimes v_1) + ac(v_1 \otimes v_2) + ac(v_2 \otimes v_1) + c^2(v_2 \otimes v_2).\end{aligned}$$

This calculation gives the first column of the matrix of $\tau \otimes \tau$ relative to the basis $v_1 \otimes v_1, v_1 \otimes v_2, v_2 \otimes v_1, v_2 \otimes v_2$. The whole matrix is

$$\begin{pmatrix} a^2 & ab & ab & b^2 \\ ac & ad & bc & bd \\ ac & bc & ad & bd \\ c^2 & cd & cd & d^2 \end{pmatrix}.$$

The map sending $\left(\begin{smallmatrix} a & b \\ c & d \end{smallmatrix}\right)$ to this matrix is a homomorphism from GL_2 to GL_4. To check (1.1.2) there is no need to make explicit matrix multiplications as in Example 1.1.5: the relation (1.1.2) follows from the fact that, for any linear transformations τ, τ' of V,

$$(\tau \circ \tau') \otimes (\tau \circ \tau') = (\tau \otimes \tau) \circ (\tau' \otimes \tau'), \tag{1.2.7}$$

which in turn follows because the two sides take the same values when evaluated on the basis elements. ∎

As can be seen in Example 1.2.3, there are two subspaces of $V \otimes V$ that are guaranteed to be preserved by all linear transformations of the form $\tau \otimes \tau$: these are the space of *symmetric tensors*, $\mathrm{Sym}^2(V)$, consisting of elements that are invariant under the interchange map $v_i \otimes v_j \mapsto v_j \otimes v_i$, and the space of *alternating tensors*, $\mathrm{Alt}^2(V)$, consisting of elements that change sign under this interchange map. By restricting $\tau \otimes \tau$ to these subspaces we obtain further homomorphisms of the type referred to in Problem 1.1.1.

Example 1.2.4. Continuing with V two-dimensional, as in Example 1.2.3, $\mathrm{Sym}^2(V)$ is three-dimensional with basis $v_1 \otimes v_1, v_1 \otimes v_2 + v_2 \otimes v_1, v_2 \otimes v_2$. The resulting homomorphism is exactly the map $\Psi : GL_2 \to GL_3$ of Example 1.1.5. By contrast, $\mathrm{Alt}^2(V)$ is one-dimensional, spanned by $v_1 \otimes v_2 - v_2 \otimes v_1$. If $\tau : V \to V$ has matrix $\left(\begin{smallmatrix} a & b \\ c & d \end{smallmatrix}\right)$ then

$$\begin{aligned}(\tau \otimes \tau)(v_1 \otimes v_2 - v_2 \otimes v_1) &= (av_1 + cv_2) \otimes (bv_1 + dv_2) \\ &\quad - (bv_1 + dv_2) \otimes (av_1 + cv_2) \\ &= (ad - bc)(v_1 \otimes v_2 - v_2 \otimes v_1).\end{aligned}$$

So the resulting homomorphism is the determinant $\det : GL_2 \to GL_1$, as in Example 1.1.2. ■

In general, if V has a basis $v_1, \dots, v_n$ then an element of $V \otimes V$ lies in $\mathrm{Sym}^2(V)$ if and only if the coefficient of $v_i \otimes v_j$ equals the coefficient of $v_j \otimes v_i$ for all i, j. Hence $\mathrm{Sym}^2(V)$ has a basis consisting of the following elements:

$$v_i \otimes v_i \text{ for } 1 \le i \le n \quad \text{and} \quad v_i \otimes v_j + v_j \otimes v_i \qquad \text{for } 1 \le i < j \le n.$$

An element of $V \otimes V$ lies in $\mathrm{Alt}^2(V)$ if and only if the coefficient of $v_i \otimes v_i$ is zero for every i and the coefficient of $v_i \otimes v_j$ is the negative of the coefficient of $v_j \otimes v_i$ for all $i \ne j$. Hence $\mathrm{Alt}^2(V)$ has a basis consisting of the elements

$$v_i \otimes v_j - v_j \otimes v_i \quad \text{for } 1 \le i < j \le n.$$

Clearly we have a direct sum decomposition,

$$V \otimes V = \mathrm{Sym}^2(V) \oplus \mathrm{Alt}^2(V), \tag{1.2.8}$$

and the dimensions of $\mathrm{Sym}^2(V)$ and $\mathrm{Alt}^2(V)$ are $\binom{n+1}{2}$ and $\binom{n}{2}$ respectively.

As well as tensor squares, one can define higher tensor powers in an entirely analogous way: $V^{\otimes 3} = V \otimes V \otimes V$, $V^{\otimes 4} = V \otimes V \otimes V \otimes V$, and so forth. If V has a basis $v_1, \dots, v_n$ then the k-fold tensor power $V^{\otimes k}$ has a basis consisting of the pure tensors:

$$v_{i_1} \otimes v_{i_2} \otimes \cdots \otimes v_{i_k} \quad \text{for } 1 \le i_1, \dots, i_k \le n.$$

So $\dim V^{\otimes k} = n^k$. (By convention, $V^{\otimes 1}$ is V itself.) The space of symmetric tensors $\mathrm{Sym}^k(V)$ consists of those elements of $V^{\otimes k}$ that are fixed under any permutation of the tensor factors. In other words, the coefficients of $v_{i_1} \otimes v_{i_2} \otimes \cdots \otimes v_{i_k}$ and $v_{j_1} \otimes v_{j_2} \otimes \cdots \otimes v_{j_k}$ have to be the same whenever $j_1, \dots, j_k$ can be obtained by rearranging $i_1, \dots, i_k$. So $\mathrm{Sym}^k(V)$ has a basis consisting of all the elements

$$t_{(k_1,\dots,k_n)} := \sum_{\substack{1 \le s_1, \dots, s_k \le n, \\ k_i \text{ of the } s_j \\ \text{equal } i}} v_{s_1} \otimes \cdots \otimes v_{s_k},$$

where $(k_1, \ldots, k_n)$ runs over all n-tuples of nonnegative integers such that $k_1 + \cdots + k_n = k$. (Note that $\mathrm{Sym}^1(V) = V$.) Hence

$$\dim \mathrm{Sym}^k(V) = \binom{n+k-1}{k}. \tag{1.2.9}$$

The space of alternating tensors $\mathrm{Alt}^k(V)$ consists of those elements of $V^{\otimes k}$ which change sign under any interchange of two tensor factors, so under a general permutation σ in the symmetric group S_k they are multiplied by the sign $\varepsilon(\sigma)$. This forces the coefficient of $v_{i_1} \otimes v_{i_2} \otimes \cdots \otimes v_{i_k}$ to be zero whenever two of the i_j coincide, and if the i_j are all distinct then it forces the coefficient of $v_{i_{\sigma(1)}} \otimes v_{i_{\sigma(2)}} \otimes \cdots \otimes v_{i_{\sigma(k)}}$ to be $\varepsilon(\sigma)$ times the coefficient of $v_{i_1} \otimes v_{i_2} \otimes \cdots \otimes v_{i_k}$, for all permutations $\sigma \in S_k$. So $\mathrm{Alt}^k(V) = \{0\}$ for $k > n$ and, for $k \leq n$, $\mathrm{Alt}^k(V)$ has a basis consisting of the elements

$$u_{i_1,\ldots,i_k} := \sum_{\sigma \in S_k} \varepsilon(\sigma)\, v_{i_{\sigma(1)}} \otimes v_{i_{\sigma(2)}} \otimes \cdots \otimes v_{i_{\sigma(k)}} \quad \text{for } 1 \leq i_1 < \cdots < i_k \leq n.$$

(Note that $\mathrm{Alt}^1(V) = V$.) Hence

$$\dim \mathrm{Alt}^k(V) = \binom{n}{k} \quad \text{(which is 0 if } k > n\text{)}. \tag{1.2.10}$$

In particular, $\mathrm{Alt}^n(V)$ is one-dimensional, spanned by the alternating tensor

$$u_{1,\ldots,n} = \sum_{\sigma \in S_n} \varepsilon(\sigma)\, v_{\sigma(1)} \otimes \cdots \otimes v_{\sigma(n)}.$$

The homomorphism $GL_n \to GL_1$ obtained by considering induced linear transformations of $\mathrm{Alt}^n(V)$ is always the determinant homomorphism, as in Example 1.2.4.

Many questions now arise. Purely from the dimensions we can see that, for $k \geq 3$, the sum of $\mathrm{Sym}^k(V)$ and $\mathrm{Alt}^k(V)$ is far from being the whole of $V^{\otimes k}$; so are there other subspaces of $V^{\otimes k}$ that are 'natural' in the sense of being preserved by any linear transformation of the form $\tau \otimes \cdots \otimes \tau$? When are the resulting homomorphisms $GL_n \to GL_m$ the same or equivalent (up to conjugation)? What do you get when you form tensor products like $\mathrm{Sym}^2(V) \otimes \mathrm{Alt}^3(V)$, or compose operations as in $\mathrm{Sym}^2(\mathrm{Alt}^3(V^*))$, or both? Such problems in multilinear algebra would seem easier to address if we had a satisfactory answer to Problem 1.1.1.

1.3 Linearization of the problem

One obstacle to solving Problem 1.1.1 directly is that the functions Φ_{ij} are in general *nonlinear*, so they cannot be specified simply by giving their values on basis elements. Lie's brilliant idea was to *linearize* Problem 1.1.1 by considering the derivative of $\Phi : GL_n \to GL_m$ at the identity matrix 1_n. By definition, this is the linear map

$$\varphi : \mathrm{Mat}_n \to \mathrm{Mat}_m$$

$$(a_{ij}) \mapsto \text{the matrix whose } (k,l) \text{ entry is } \sum_{i,j=1}^{n} a_{ij}(\partial^{ij}\Phi_{kl})(1_n),$$

where ∂^{ij} denotes the partial derivative with respect to the (i,j)-entry. To make the idea work we obviously need these partial derivatives to exist, so this condition should be retrospectively added to Problem 1.1.1. Certainly the partial derivatives exist in all the examples we have seen so far, where the functions Φ_{kl} were polynomials or rational functions in the entries.

Example 1.3.1. The derivative of the homomorphism $\Psi : GL_2 \to GL_3$ in Example 1.1.5 is the linear map $\psi : \mathfrak{gl}_2 \to \mathfrak{gl}_3$ given by

$$\psi\begin{pmatrix} a & b \\ c & d \end{pmatrix} = \begin{pmatrix} 2a & 2b & 0 \\ c & a+d & b \\ 0 & 2c & 2d \end{pmatrix}.$$

One might wonder whether ψ satisfies (1.1.2). To find out, we make the following calculation:

$$\psi\begin{pmatrix} a & b \\ c & d \end{pmatrix}\psi\begin{pmatrix} e & f \\ g & h \end{pmatrix} - \psi\left(\begin{pmatrix} a & b \\ c & d \end{pmatrix}\begin{pmatrix} e & f \\ g & h \end{pmatrix}\right)$$
$$= \begin{pmatrix} 2a & 2b & 0 \\ c & a+d & b \\ 0 & 2c & 2d \end{pmatrix}\begin{pmatrix} 2e & 2f & 0 \\ g & e+h & f \\ 0 & 2g & 2h \end{pmatrix}$$
$$- \begin{pmatrix} 2(ae+bg) & 2(af+bh) & 0 \\ ce+dg & (ae+bg)+(cf+dh) & af+bh \\ 0 & 2(ce+dg) & 2(cf+dh) \end{pmatrix}$$
$$= \begin{pmatrix} 2ae & 2af+2be & 2bf \\ ce+ag & ah+bg+cf+de & df+bh \\ 2cg & 2ch+2dg & 2dh \end{pmatrix}.$$

The result is not zero, but it is noticeable that it is symmetric in the two matrices used (it remains unchanged if we swap a with e, b with f, c with g, and d with h). In other words, ψ does satisfy the property

$$\psi(x)\psi(y) - \psi(xy) = \psi(y)\psi(x) - \psi(yx) \quad \text{for all } x, y \in \mathrm{Mat}_2.$$

■

Lie observed that this identity holds in general.

Proposition 1.3.2. *If $\Phi : GL_n \to GL_m$ is a group homomorphism such that all the partial derivatives $\partial^{ij}\Phi_{kl}$ exist at 1_n, and $\varphi : \mathrm{Mat}_n \to \mathrm{Mat}_m$ is the derivative of Φ at 1_n as above, then*

$$\varphi(xy - yx) = \varphi(x)\varphi(y) - \varphi(y)\varphi(x) \quad \textit{for all } x, y \in \mathrm{Mat}_n.$$

Proof. (The reader who is prepared to accept the result can safely skip this proof; it relies on moderately sophisticated calculus techniques which we will not need elsewhere.) Since Φ is a homomorphism, we have

$$\Phi(ghg^{-1}) = \Phi(g)\Phi(h)\Phi(g)^{-1} \quad \text{for all } g, h \in GL_n. \tag{1.3.1}$$

The first step is to differentiate both sides of this equation as functions of h, keeping g fixed, and evaluating the result at $h = 1_n$. For all $g \in GL_n$, the conjugation map $GL_n \to GL_n : h \mapsto ghg^{-1}$ is the restriction of the linear map $\mathrm{Mat}_n \to \mathrm{Mat}_n : y \mapsto gyg^{-1}$, so its derivative at 1_n is this linear map. By the chain rule the derivative of $GL_n \to GL_m : h \mapsto \Phi(ghg^{-1})$ at 1_n is $\mathrm{Mat}_n \to \mathrm{Mat}_m : y \mapsto \varphi(gyg^{-1})$. Reasoning similarly for the right-hand side of (1.3.1), we obtain the desired differentiated equation:

$$\varphi(gyg^{-1}) = \Phi(g)\varphi(y)\Phi(g)^{-1} \quad \text{for all } g \in GL_n, y \in \mathrm{Mat}_n. \tag{1.3.2}$$

The remaining step is to differentiate both sides again but this time as functions of g, keeping y fixed and evaluating the result at $g = 1_n$. For any $y \in \mathrm{Mat}_n$, the derivative of $GL_n \to \mathrm{Mat}_n : g \mapsto gyg^{-1}$ at 1_n is $\mathrm{Mat}_n \to \mathrm{Mat}_n : x \mapsto xy - yx$, by virtue of the power series expansion

$$\begin{aligned}(1_n + tx)y(1_n + tx)^{-1} &= \sum_{i=0}^{\infty}(1_n + tx)y(-tx)^i \\ &= y + t(xy - yx) + t^2(yx^2 - xyx) + \cdots .\end{aligned} \tag{1.3.3}$$

So, again by the chain rule, the derivative of $GL_n \to \mathrm{Mat}_m : g \mapsto \varphi(gyg^{-1})$ at 1_n is $\mathrm{Mat}_n \to \mathrm{Mat}_m : x \mapsto \varphi(xy - yx)$. By similar reasoning, the derivative of $GL_n \to \mathrm{Mat}_m : g \mapsto \Phi(g)\varphi(y)\Phi(g)^{-1}$ at 1_n is $\mathrm{Mat}_n \to \mathrm{Mat}_m : x \mapsto \varphi(x)\varphi(y) - \varphi(y)\varphi(x)$. So the differentiation of (1.3.2) gives the result. □

Proposition 1.3.2 can be written more concisely using the bracket notation $[x, y] := xy - yx$. It says that

$$\varphi([x, y]) = [\varphi(x), \varphi(y)] \quad \text{for all } x, y \in \mathrm{Mat}_n. \tag{1.3.4}$$

The quantity $[x, y]$ is called the *commutator* of x and y, since it measures to what extent the commutation equation $xy = yx$ fails to hold.

We are now in the curious situation of being interested in linear maps $\varphi : \mathrm{Mat}_n \to \mathrm{Mat}_m$ that preserve the commutator, in the sense of (1.3.4), but do not necessarily preserve the actual multiplication. The structure on Mat_n given solely by the commutator is that of a *Lie algebra* and is our first example of the topic of this book. When Mat_n is viewed as a Lie algebra (i.e. when matrix multiplication is not under consideration, but commutators are) it is written as $\mathfrak{gl}_n$; the notation is meant to suggest that it is a linearized version of the group GL_n. Linear maps $\varphi : \mathfrak{gl}_n \to \mathfrak{gl}_m$ satisfying (1.3.4) are called *Lie algebra homomorphisms*. So the linearization of Problem 1.1.1 is the following.

Problem 1.3.3. Describe all Lie algebra homomorphisms $\varphi : \mathfrak{gl}_n \to \mathfrak{gl}_m$.

Mainly because it concerns linear maps, Problem 1.3.3 is easier to tackle than Problem 1.1.1. Nevertheless, it is not a trivial matter: much of the rest of this book will be devoted to refining the statement of Problem 1.3.3 and solving it.

1.4 Lie's theorem

It may now appear that we have carried out the customary mathematician's dodge of replacing the problem that we really intended to solve, Problem 1.1.1, with a problem that is more tractable, Problem 1.3.3. However, because of the bridge between Lie groups and Lie algebras the two problems are almost equivalent. In this section we will briefly explain why this is so.

The first key result is that the derivative φ determines the homomorphism Φ. This is surprising, because you would think there could be two homomorphisms $GL_n \to GL_m$ that had the same derivative at 1_n but different 'higher-order terms'. We need to use the fact that every element of GL_n can be written as $\exp(x)$ for some $x \in \mathrm{Mat}_n$, where exp is the matrix exponential defined by the (always convergent) series

$$\exp(x) = 1_n + x + \frac{x^2}{2!} + \frac{x^3}{3!} + \cdots .$$

Proposition 1.4.1. *Under the hypotheses of Proposition 1.3.2,*

$$\Phi(\exp(x)) = \exp(\varphi(x)) \quad \textit{for all } x \in \mathrm{Mat}_n.$$

So Φ is determined by φ.

Proof. Consider the Mat_m-valued functions of one variable

$$f(t) = \Phi(\exp(tx)) \quad \text{and} \quad g(t) = \exp(-\varphi(tx)).$$

We have $f(t+s) = f(t)f(s)$ and $g(s+t) = g(s)g(t)$. It follows that

$$f'(t) = \lim_{s\to 0} \frac{f(t+s) - f(t)}{s} = f(t) \lim_{s\to 0} \frac{f(s) - f(0)}{s} = f(t)\varphi(x)$$

and similarly $g'(t) = -\varphi(x)g(t)$. Hence

$$(f(t)g(t))' = f'(t)g(t) + f(t)g'(t) = f(t)\varphi(x)g(t) - f(t)\varphi(x)g(t) = 0,$$

which means that $f(t)g(t)$ is the constant function with value $f(0)g(0) = 1_m$. In particular, $f(1) = g(1)^{-1}$, which is the claim. □

Another important question is whether every Lie algebra homomorphism $\varphi : \mathfrak{gl}_n \to \mathfrak{gl}_m$ can be 'integrated', i.e. whether it arises as the derivative of a group homomorphism $\Phi : GL_n \to GL_m$. We can show that the answer is no. Let e_{ij} be the matrix with 1 in the (i, j)-position and zeroes elsewhere. Then, since $\exp(2\pi\mathbf{i}e_{ii}) = 1_n$ for all i, it follows from Proposition 1.4.1 that $\exp(2\pi\mathbf{i}\varphi(e_{ii})) = 1_m$. Calculating this exponential using the Jordan canonical form, one deduces that, under the hypotheses of Proposition 1.3.2,

$$\varphi(e_{ii}) \text{ is diagonalizable with integer eigenvalues, for all } i. \tag{1.4.1}$$

This is not a property possessed by an arbitrary Lie algebra homomorphism $\varphi : \mathfrak{gl}_n \to \mathfrak{gl}_m$: for instance, when $n = m = 1$, any scalar multiplication $\mathbb{C} \to \mathbb{C}$ is a Lie algebra homomorphism but only those where the scalar is an integer satisfy (1.4.1). So we should modify Problem 1.3.3 to the following:

Problem 1.4.2. Describe all Lie algebra homomorphisms $\varphi : \mathfrak{gl}_n \to \mathfrak{gl}_m$ such that (1.4.1) holds.

It turns out that Problem 1.4.2 is equivalent to Problem 1.1.1. This is still somewhat surprising, because it seems likely that there would be other integrability conditions besides (1.4.1): for instance, there are many matrices x other than e_{ii} for which we can deduce a similar constraint on $\varphi(x)$. Also, even if φ is such that a map $\Phi : GL_n \to GL_m$ is well defined by the formula in Proposition 1.4.1, there would seem to be no guarantee that Φ satisfies (1.1.2); recall that $\exp(x + y)$ does not equal $\exp(x)\exp(y)$ in general (for example, take $x = e_{12}$ and $y = e_{21}$).

However, it follows from Lie's results that any Lie algebra homomorphism $\varphi : \mathfrak{gl}_n \to \mathfrak{gl}_m$ satisfying (1.4.1) can be integrated to a group homomorphism $\Phi : GL_n \to GL_m$. We will be able to give more explanation of this after solving Problem 1.4.2; see Remark 7.5.6.

Although in this chapter we have concentrated on GL_n, the concepts apply to much more general Lie groups. In particular, for any subgroup G of GL_n (satisfying the topological assumptions needed to qualify as a Lie group, which we will not discuss here) there is a corresponding Lie subalgebra $\mathfrak{g}$ of $\mathfrak{gl}_n$, i.e. a subspace that is closed under the operation of taking commutators but not necessarily closed under multiplication: by definition, $\mathfrak{g}$ is the *tangent space* to G at the identity 1_n.

Example 1.4.3. An important subgroup of GL_n is the *special linear group*

$$SL_n = \{g \in GL_n \mid \det(g) = 1\}.$$

Using the fact that the derivative of $\det : GL_n \to \mathbb{C}^\times$ at 1_n is the trace map $\mathrm{tr} : \mathrm{Mat}_n \to \mathbb{C}$, one finds that the corresponding Lie algebra is

$$\mathfrak{sl}_n = \{x \in \mathrm{Mat}_n \mid \mathrm{tr}(x) = 0\},$$

consisting of all the trace-zero matrices. ∎

Example 1.4.4. Another subgroup of GL_n is the *special orthogonal group*

$$SO_n = \{g \in SL_n \mid g^{\mathrm{t}} = g^{-1}\}.$$

Since the derivative of the inversion map $GL_n \to GL_n : g \mapsto g^{-1}$ at 1_n is the map $x \mapsto -x$, it turns out that the corresponding Lie algebra is

$$\mathfrak{so}_n = \{x \in \mathrm{Mat}_n \mid x^{\mathrm{t}} = -x\},$$

consisting of all the skew-symmetric matrices. ∎

As in the case of the general linear group, suitably differentiable group homomorphisms $\Phi : G \to H$ between arbitrary Lie groups can be studied in terms of their derivatives, which are Lie algebra homomorphisms $\varphi : \mathfrak{g} \to \mathfrak{h}$. The precise statement, known as Lie's theorem, provides a fairly complete 'dictionary' for translating statements about Lie groups into statements about Lie algebras; this allows a wide range of difficult problems to be linearized and solved. For a proper treatment of Lie's theorem, see (for instance) [3], [8], [13], [15], or the ultimate reference [1].

In the main part of the book we will study Lie algebras from a completely algebraic point of view, without referring back to Lie groups (except for occasional side-remarks). Unfortunately, this means that we will be omitting some leading applications of the theory including applications to differential equations, which were Lie's original motivation. The advantage is that we can evade the technicalities of differential geometry and harmonic analysis and still go reasonably far into one of the most beautiful and rich bodies of theory that mathematics has to offer.

CHAPTER 2

Definition of a Lie algebra

As mentioned in the previous chapter, the first examples of Lie algebras were vector spaces of matrices equipped with a 'multiplication', which was not the usual matrix multiplication but the commutator $[x, y] = xy - yx$. The commutator seems to be a fairly defective kind of multiplication, since it is not associative. But, in measuring the failure of associativity, we discover an interesting fact:

$$\begin{aligned}[x,[y,z]] - [[x,y],z] &= x(yz - zy) - (yz - zy)x - (xy - yx)z + z(xy - yx) \\ &= xyz - xzy - yzx + zyx - xyz + yxz + zxy - zyx \\ &= y(xz - zx) - (xz - zx)y \\ &= [y,[x,z]].\end{aligned}$$

Thus we are led to consider algebraic operations $[x, y]$ that satisfy this *Jacobi identity*, $[x,[y,z]] = [[x,y],z] + [y,[x,z]]$, without necessarily arising as a commutator.

2.1 Definition and first examples

Here are the axioms with which the theory of Lie algebras begins.

Definition 2.1.1. A *Lie algebra* is a vector space $\mathfrak{g}$ with a *Lie bracket* map $\mathfrak{g} \times \mathfrak{g} \to \mathfrak{g}$, written $(x, y) \mapsto [x, y]$, satisfying:

(1) the Lie bracket is bilinear, i.e., for all $a, b \in \mathbb{C}$ and $x, y, z \in \mathfrak{g}$,

$$\begin{aligned}[ax + by, z] &= a[x, z] + b[y, z], \\ [x, ay + bz] &= a[x, y] + b[x, z];\end{aligned}$$

(2) the Lie bracket is skew-symmetric, i.e. $[y, x] = -[x, y]$ for all $x, y \in \mathfrak{g}$;

(3) the Lie bracket satisfies the Jacobi identity, i.e.

$$[x,[y,z]] = [[x,y],z] + [y,[x,z]] \quad \text{for all } x, y, z \in \mathfrak{g}.$$

Note that (2) implies $[x, x] = 0$ for all $x \in \mathfrak{g}$. Using (1) and (2) we can obtain many equivalent formulations of (3), for instance

$$[[x,y],z] = [[x,z],y] + [x,[y,z]]$$

and

$$[x, [y, z]] + [y, [z, x]] + [z, [x, y]] = 0.$$

The latter form is easy to remember because x, y, z are cyclically permuted in the three terms. Notice that we do not get anything interesting by choosing two variables in the Jacobi identity to be the same: for example, if we set $x = y$ then the $[[x, y], z]$ term vanishes because $[x, x] = 0$ and the equation becomes obvious.

We can instantly identify some broad classes of examples of Lie algebras:

Proposition 2.1.2. *Any associative $\mathbb{C}$-algebra A becomes a Lie algebra if we use the commutator $[x, y] = xy - yx$ as the Lie bracket.*

Proof. The Jacobi identity was checked above. Bilinearity follows from the bilinearity of multiplication in A, and skew-symmetry is obvious from the definition. □

Corollary 2.1.3. *Any subspace of an associative algebra that is closed under the taking of commutators is a Lie algebra, with the commutator as the Lie bracket.*

Proof. Obviously the three axioms remain true when the Lie bracket is restricted to the subspace. □

Definition 2.1.4. A *linear Lie algebra* is a subspace of Mat_n for some n that is closed under the taking of commutators.

Example 2.1.5. The three classes of linear Lie algebras mentioned in the previous chapter are the *general linear Lie algebra*

$$\mathfrak{gl}_n = \mathrm{Mat}_n \text{ (considered as a Lie algebra)},$$

the *special linear Lie algebra*

$$\mathfrak{sl}_n = \{x \in \mathrm{Mat}_n \mid \mathrm{tr}(x) = 0\},$$

and the *special orthogonal Lie algebra*

$$\mathfrak{so}_n = \{x \in \mathrm{Mat}_n \mid x^{\mathsf{t}} = -x\}.$$

It is clear that $\mathfrak{sl}_n$ and $\mathfrak{so}_n$ are subspaces of Mat_n. The reason that $\mathfrak{sl}_n$ is closed under the taking of commutators is that, for any $x, y \in \mathrm{Mat}_n$ (and, in particular, for $x, y \in \mathfrak{sl}_n$),

$$\mathrm{tr}([x, y]) = \mathrm{tr}(xy) - \mathrm{tr}(yx) = 0, \tag{2.1.1}$$

by a well-known property of the trace function. To show that $\mathfrak{so}_n$ is closed under commutators, let $x, y \in \mathfrak{so}_n$ and observe that

$$\begin{aligned}[x, y]^{\mathsf{t}} &= (xy - yx)^{\mathsf{t}} = y^{\mathsf{t}}x^{\mathsf{t}} - x^{\mathsf{t}}y^{\mathsf{t}} \\ &= (-y)(-x) - (-x)(-y) = yx - xy = -[x, y],\end{aligned}$$

which means that $[x, y] \in \mathfrak{so}_n$. The dimensions of these Lie algebras are as follows:

$$\dim \mathfrak{gl}_n = n^2, \quad \dim \mathfrak{sl}_n = n^2 - 1, \quad \dim \mathfrak{so}_n = \binom{n}{2}. \tag{2.1.2}$$

To explain the expression for $\dim \mathfrak{so}_n$, observe that in any skew-symmetric matrix the diagonal entries must all be 0 and the below-diagonal entries are determined by the above-diagonal entries. We will identify $\mathfrak{gl}_1 = \mathrm{Mat}_1$ with $\mathbb{C}$ in an obvious way; note that $\mathfrak{sl}_1$ and $\mathfrak{so}_1$ are both $\{0\}$. ■

Here is a rather uninteresting way of defining a Lie algebra:

Proposition 2.1.6. *Any vector space V becomes a Lie algebra if we set all the Lie brackets equal to* 0.

Proof. All three axioms are trivially true. □

Definition 2.1.7. A Lie algebra $\mathfrak{g}$ with $[x, y] = 0$ for all $x, y \in \mathfrak{g}$ is said to be *abelian*.

Example 2.1.8. If you apply Proposition 2.1.2 to a commutative associative algebra, you get an abelian Lie algebra because all the commutators are zero. ■

It is common to say that two elements x and y of a Lie algebra *commute* if $[x, y] = 0$. If the Lie bracket is the commutator in an associative algebra as in Proposition 2.1.2, this equation is equivalent to the usual equation $xy = yx$, but, for a general Lie algebra, the expressions xy and yx have no meaning.

Remark 2.1.9. If G is an abelian subgroup of GL_n that is a Lie group, the corresponding Lie algebra (see Section 1.4) is abelian.

Proposition 2.1.10. *Any one-dimensional Lie algebra is abelian.*

Proof. Let $\mathfrak{g} = \mathbb{C}x$ be a one-dimensional Lie algebra spanned by x. Then for all $a, b \in \mathbb{C}$, we have $[ax, bx] = ab[x, x] = 0$ by skew-symmetry. □

2.2 Classification and isomorphisms

When specifying the Lie bracket on a Lie algebra with basis $x_1, \ldots, x_m$, it is enough to specify the Lie brackets $[x_i, x_j]$, where $i < j$. The Lie bracket of two general elements is then given by

$$\begin{aligned}
[a_1x_1 + \cdots + a_mx_m, b_1x_1 + \cdots + b_mx_m] &= \sum_{i,j=1}^{m} a_ib_j[x_i, x_j] \\
&= \sum_{i<j} (a_ib_j - a_jb_i)[x_i, x_j], \qquad (2.2.1)
\end{aligned}$$

where the first equation holds by bilinearity and the second by skew-symmetry.

Since the Jacobi identity (see the start of the chapter) is linear in each of the three elements involved, if it holds whenever those three elements belong to some chosen basis then it must hold in general.

Example 2.2.1. Suppose that $[\cdot,\cdot]$ is a bracket on a two-dimensional vector space $\mathfrak{g}$ that is bilinear and skew-symmetric. The Jacobi identity holds automatically when the three elements appearing in it belong to a chosen basis of $\mathfrak{g}$ because at least two of the three must coincide, making the equation trivial. So axiom (3) of Definition 2.2.1 is redundant if $\mathfrak{g}$ is two-dimensional. ■

Example 2.2.2. Let $\mathfrak{l}_2$ be a two-dimensional vector space with basis s, t. If we specify that $[s,t] = t$ then skew-symmetry forces $[s,s] = [t,t] = 0$ and $[t,s] = -t$, and bilinearity forces the general definition

$$[as + bt, cs + dt] = (ad - bc)t \quad \text{for all } a, b, c, d \in \mathbb{C}.$$

By Example 2.2.1 this defines a Lie algebra. ■

When making calculations in $\mathfrak{gl}_n$, we will normally use the standard basis $e_{11}, e_{12}, \ldots, e_{1n}, e_{21}, \ldots, e_{nn}$, where, as mentioned earlier, e_{ij} is the matrix with 1 in the (i,j)-position and zeroes everywhere else. Matrix multiplication is defined by

$$e_{ij}e_{kl} = \delta_{jk}e_{il}, \tag{2.2.2}$$

where as usual δ_{jk} is 1 if $j = k$ and 0 otherwise. So the Lie bracket on $\mathfrak{gl}_n$ can be defined by the rule

$$[e_{ij}, e_{kl}] = \delta_{jk}e_{il} - \delta_{li}e_{kj}; \tag{2.2.3}$$

this rule can be extended to arbitrary elements by bilinearity.

Example 2.2.3. In $\mathfrak{gl}_2$, the standard basis consists of

$$e_{11} = \begin{pmatrix} 1 & 0 \\ 0 & 0 \end{pmatrix}, \; e_{12} = \begin{pmatrix} 0 & 1 \\ 0 & 0 \end{pmatrix}, \; e_{21} = \begin{pmatrix} 0 & 0 \\ 1 & 0 \end{pmatrix}, \; e_{22} = \begin{pmatrix} 0 & 0 \\ 0 & 1 \end{pmatrix}$$

and the Lie bracket is completely determined by

$$\begin{aligned} &[e_{11}, e_{12}] = e_{12}, \quad [e_{11}, e_{21}] = -e_{21}, \quad [e_{11}, e_{22}] = 0, \\ &[e_{12}, e_{21}] = e_{11} - e_{22}, \quad [e_{12}, e_{22}] = e_{12}, \quad [e_{21}, e_{22}] = -e_{21}. \end{aligned}$$

One can see immediately that many spans of subsets of the basis are closed under taking commutators: to give two examples, the diagonal subspace

$$\mathbb{C}\{e_{11}, e_{22}\} = \begin{pmatrix} * & 0 \\ 0 & * \end{pmatrix}$$

is a two-dimensional abelian Lie algebra, and the top-row subspace

$$\mathbb{C}\{e_{11}, e_{12}\} = \begin{pmatrix} * & * \\ 0 & 0 \end{pmatrix}$$

is a two-dimensional non-abelian Lie algebra. Indeed, the latter has the bracket relation $[e_{11}, e_{12}] = e_{12}$, identical to that of $\mathfrak{l}_2$ except that s has been renamed e_{11} and t has been renamed e_{12}. Thus $\mathbb{C}\{e_{11}, e_{12}\}$ and $\mathfrak{l}_2$ are *isomorphic* Lie algebras in the sense of the definition below. ■

As with groups and rings, we define a homomorphism to be a map that preserves the relevant algebraic structure:

Definition 2.2.4. If $\mathfrak{g}$ and $\mathfrak{g}'$ are Lie algebras, a *homomorphism* $\varphi : \mathfrak{g} \to \mathfrak{g}'$ is a linear map such that

$$\varphi([x, y]) = [\varphi(x), \varphi(y)] \quad \text{for all } x, y \in \mathfrak{g}.$$

As usual, the identity map is a homomorphism, and the composition of two homomorphisms is another homomorphism (so Lie algebras and the homomorphisms between them form a *category*). For any two Lie algebras $\mathfrak{g}$ and $\mathfrak{g}'$, the zero map $\mathfrak{g} \to \mathfrak{g}' : x \mapsto 0$ is trivially a homomorphism. Note, however, that the homomorphism condition is not linear in φ: the set of all homomorphisms from $\mathfrak{g}$ to $\mathfrak{g}'$ is not, in general, a subspace of the vector space of linear maps $\mathfrak{g} \to \mathfrak{g}'$.

If $\mathfrak{g}$ has a basis $x_1, \ldots, x_m$ then the condition for a map φ to be a homomorphism reduces to the case that $x = x_i$ and $y = x_j$ where $i < j$, because the case $i > j$ then follows by skew-symmetry, the case $i = j$ holds trivially, and the general case follows by bilinearity.

Definition 2.2.5. A bijective homomorphism is an *isomorphism* (its inverse is clearly also an isomorphism). We write $\mathfrak{g} \cong \mathfrak{g}'$ to mean that $\mathfrak{g}$ and $\mathfrak{g}'$ are *isomorphic*, i.e. there exists an isomorphism between them.

Isomorphic Lie algebras are, for almost all purposes, the same. It is immediate from the definition that two Lie algebras are isomorphic if and only if one can find bases for them with respect to which the Lie brackets are given by the same formulas, as in the case of $\mathfrak{l}_2$ and $\mathbb{C}\{e_{11}, e_{12}\}$.

Example 2.2.6. Obviously, all abelian Lie algebras of a fixed dimension are isomorphic. In particular, there is only a single one-dimensional Lie algebra up to isomorphism; we will typically call it just $\mathbb{C}$ (or $\mathbb{C}x$ when we need to indicate a basis element x). For instance, $\mathfrak{so}_2 \cong \mathbb{C}$. ■

Proposition 2.2.7. *Any two-dimensional Lie algebra is either abelian or is isomorphic to* $\mathfrak{l}_2$.

Proof. Let $\mathfrak{g}$ be a two-dimensional non-abelian Lie algebra with basis x, y. We need to find another basis u, v of $\mathfrak{g}$ such that $[u, v] = v$. Suppose that $[x, y] = ax + by$. At least one of a and b is nonzero since otherwise $\mathfrak{g}$ would be abelian. If $b \neq 0$, let $u = b^{-1}x$ and $v = ax + by$. Then

$$[u, v] = b^{-1}[x, ax + by] = [x, y] = ax + by = v.$$

If $b = 0$, let $u = -a^{-1}y$, $v = x$. Then

$$[u, v] = a^{-1}[x, y] = a^{-1}ax = v.$$

So, in either case we are done. □

Classifying three-dimensional Lie algebras up to isomorphism is more difficult, especially from this naive point of view; one needs a more structural approach (see Exercises 2.3.8, 3.4.6, 3.4.7, and 4.5.7). We can, however, examine the most important three-dimensional Lie algebra, namely $\mathfrak{sl}_2$, consisting of all 2×2 matrices with trace 0. The standard basis of $\mathfrak{sl}_2$ is e, h, f, where

$$e = e_{12} = \begin{pmatrix} 0 & 1 \\ 0 & 0 \end{pmatrix}, \quad h = e_{11} - e_{22} = \begin{pmatrix} 1 & 0 \\ 0 & -1 \end{pmatrix}, \quad f = e_{21} = \begin{pmatrix} 0 & 0 \\ 1 & 0 \end{pmatrix}.$$

The Lie bracket is completely specified by the brackets of these basis elements, which are given by:

Proposition 2.2.8. $[e, f] = h$, $[h, e] = 2e$, $[h, f] = -2f$.

Proof. Using the formulas in Example 2.2.3, we get

$$\begin{aligned}
[e, f] &= [e_{12}, e_{21}] = e_{11} - e_{22} = h, \\
[h, e] &= [e_{11}, e_{12}] - [e_{22}, e_{12}] = e_{12} + e_{12} = 2e, \\
[h, f] &= [e_{11}, e_{21}] - [e_{22}, e_{21}] = -e_{21} - e_{21} = -2f,
\end{aligned}$$

as claimed. □

2.3 Exercises

Exercise 2.3.1. Suppose that $\mathfrak{g}$ is a three-dimensional Lie algebra with basis x_1, x_2, x_3 such that

$$\begin{aligned}
[x_1, x_2] &= x_1 + x_2, \\
[x_1, x_3] &= ax_1 + x_3, \\
[x_2, x_3] &= x_2 + bx_3,
\end{aligned}$$

for some $a, b \in \mathbb{C}$. Determine a and b.

Exercise 2.3.2. Let $\mathfrak{g}$ be a vector space and choose a nonzero element $v \in \mathfrak{g}$. Given a function $f : \mathfrak{g} \times \mathfrak{g} \to \mathbb{C}$, we can define a map $[\cdot,\cdot] : \mathfrak{g} \times \mathfrak{g} \to \mathfrak{g}$ by $[x, y] = f(x, y)v$. What conditions must f and v satisfy in order for this to give a Lie algebra?

Exercise 2.3.3. Show that if $\mathfrak{g}$ is an abelian Lie algebra then the only homomorphism from $\mathfrak{sl}_2$ to $\mathfrak{g}$ is the zero map. Is there a nonzero homomorphism from an abelian Lie algebra to $\mathfrak{sl}_2$?

Exercise 2.3.4. Continuing Example 2.2.3, find all subsets of the standard basis of $\mathfrak{gl}_2$ whose span is closed under the taking of commutators. Classify the resulting Lie algebras up to isomorphism.

Exercise 2.3.5. The Lie algebra $\mathfrak{so}_3$ is three dimensional. Its most obvious basis consists of

$$x = \begin{pmatrix} 0 & 1 & 0 \\ -1 & 0 & 0 \\ 0 & 0 & 0 \end{pmatrix}, \quad y = \begin{pmatrix} 0 & 0 & 0 \\ 0 & 0 & 1 \\ 0 & -1 & 0 \end{pmatrix}, \quad z = \begin{pmatrix} 0 & 0 & 1 \\ 0 & 0 & 0 \\ -1 & 0 & 0 \end{pmatrix}.$$

(i) Find the Lie brackets of these basis elements.
(ii) Find an isomorphism $\varphi : \mathfrak{sl}_2 \xrightarrow{\sim} \mathfrak{so}_3$. This amounts to finding another basis $\varphi(e), \varphi(h), \varphi(f)$ of $\mathfrak{so}_3$ that satisfies the $\mathfrak{sl}_2$ bracket relations. (*Hint*: there is a solution in which $\varphi(h)$ is a multiple of x.)
(iii) Recall the Lie algebra homomorphism $\psi : \mathfrak{gl}_2 \to \mathfrak{gl}_3$ from Example 1.3.1. Find an invertible matrix $g \in GL_3$ such that the isomorphism you found in the previous part is given by $u \mapsto g\psi(u)g^{-1}$.

Exercise 2.3.6. Show that none of the Lie algebras appearing in Exercise 2.3.4 is isomorphic to $\mathfrak{sl}_2$.

Exercise 2.3.7. An isomorphism from a Lie algebra $\mathfrak{g}$ to itself is called an *automorphism* of $\mathfrak{g}$.

(i) Find all automorphisms of $\mathfrak{l}_2$.
(ii) Show that, for all $g \in GL_n$, the conjugation map $x \mapsto gxg^{-1}$ is an automorphism of $\mathfrak{gl}_n$.
(iii) Show that the map $x \mapsto -x^{\mathrm{t}}$ is an automorphism of $\mathfrak{gl}_n$.
(iv) Show that every automorphism of $\mathfrak{sl}_2$ is the restriction of one of the conjugation automorphisms of $\mathfrak{gl}_2$ defined in (ii). (*Hint*: the conjugation automorphisms form a group under composition. So, in proving that any automorphism τ of $\mathfrak{sl}_2$ is a conjugation automorphism, we are allowed to compose

τ with a conjugation automorphism. Thus we can assume that $\tau(h)$ is in Jordan canonical form.)

Exercise 2.3.8. For any $\left(\begin{smallmatrix} a & b \\ c & d \end{smallmatrix}\right) \in GL_2$, define a vector space $\mathfrak{g}_{\left(\begin{smallmatrix} a & b \\ c & d \end{smallmatrix}\right)}$ with basis x, y, z and a bilinear skew-symmetric bracket $[\cdot, \cdot]$ given by

$$[x, y] = ay + cz,$$
$$[x, z] = by + dz,$$
$$[y, z] = 0.$$

(i) Show that $\mathfrak{g}_{\left(\begin{smallmatrix} a & b \\ c & d \end{smallmatrix}\right)}$ is a Lie algebra (i.e. check the Jacobi identity).

(ii) Show that $\mathfrak{g}_{\left(\begin{smallmatrix} a & b \\ c & d \end{smallmatrix}\right)} \cong \mathfrak{g}_{\left(\begin{smallmatrix} a' & b' \\ c' & d' \end{smallmatrix}\right)}$ if and only if $\left(\begin{smallmatrix} a' & b' \\ c' & d' \end{smallmatrix}\right) = \lambda g \left(\begin{smallmatrix} a & b \\ c & d \end{smallmatrix}\right) g^{-1}$ for some $\lambda \in \mathbb{C}^\times$ and $g \in GL_2$.

(iii) Hence show that $\mathfrak{g}_{\left(\begin{smallmatrix} a & b \\ c & d \end{smallmatrix}\right)}$ is isomorphic either to $\mathfrak{g}_{\left(\begin{smallmatrix} 1 & 1 \\ 0 & 1 \end{smallmatrix}\right)}$ or to $\mathfrak{g}_{\left(\begin{smallmatrix} \alpha & 0 \\ 0 & 1 \end{smallmatrix}\right)}$ for $\alpha \in \mathbb{C}^\times$ and that in the latter case the pair $\{\alpha, \alpha^{-1}\}$ is uniquely determined.

Thus we have classified all three-dimensional Lie algebras that have a Lie bracket of the above special form with respect to some basis. Does the Lie algebra in Exercise 2.3.1 fit into this classification?

CHAPTER 3

Basic structure of a Lie algebra

In the case of algebraic structures such as groups and rings, there are basic results about 'internal' features (subgroups, subrings) and their relationship with 'external' features (such as homomorphisms) that need to be proved before any significant progress can be made. The theory of Lie algebras follows much the same pattern.

3.1 Lie subalgebras

The first necessary definition is that of the analogue of a subgroup or subring.

Definition 3.1.1. A *Lie subalgebra* (usually just *subalgebra*) of a Lie algebra $\mathfrak{g}$ is a subspace $\mathfrak{h}$ that is closed under the Lie bracket, i.e. that satisfies

$$x, y \in \mathfrak{h} \Rightarrow [x, y] \in \mathfrak{h}.$$

In the previous chapter we met this concept implicitly: linear Lie algebras are precisely the Lie subalgebras of $\mathfrak{gl}_n$ for various n. As in that case, we have:

Proposition 3.1.2. *Any subalgebra $\mathfrak{h}$ of a Lie algebra $\mathfrak{g}$ is a Lie algebra in its own right, where the Lie bracket is the restriction of that of $\mathfrak{g}$.*

Proof. This is obvious. □

Example 3.1.3. Some important subalgebras of $\mathfrak{gl}_n$ are the following spans of subsets of the standard basis: the diagonal subalgebra

$$\mathfrak{d}_n = \left\{ \begin{pmatrix} * & 0 & \cdots & 0 & 0 \\ 0 & * & \cdots & 0 & 0 \\ \vdots & \vdots & \ddots & \vdots & \vdots \\ 0 & 0 & \cdots & * & 0 \\ 0 & 0 & \cdots & 0 & * \end{pmatrix} \right\} = \mathbb{C}\{e_{ii} \,|\, 1 \le i \le n\};$$

the upper-triangular subalgebra

$$\mathfrak{b}_n = \left\{ \begin{pmatrix} * & * & \cdots & * & * \\ 0 & * & \cdots & * & * \\ \vdots & \vdots & \ddots & \vdots & \vdots \\ 0 & 0 & \cdots & * & * \\ 0 & 0 & \cdots & 0 & * \end{pmatrix} \right\} = \mathbb{C}\{e_{ij} \,|\, 1 \le i \le j \le n\};$$

and the strictly upper-triangular subalgebra

$$\mathfrak{n}_n = \left\{ \begin{pmatrix} 0 & * & \cdots & * & * \\ 0 & 0 & \cdots & * & * \\ \vdots & \vdots & \ddots & \vdots & \vdots \\ 0 & 0 & \cdots & 0 & * \\ 0 & 0 & \cdots & 0 & 0 \end{pmatrix} \right\} = \mathbb{C}\{e_{ij} \,|\, 1 \le i < j \le n\}.$$

In fact, it is easy to see that these subalgebras are closed under matrix multiplication (i.e. they are subalgebras of Mat_n), not just closed under the taking of commutators. The dimensions of these Lie algebras are as follows:

$$\dim \mathfrak{d}_n = n, \quad \dim \mathfrak{b}_n = \binom{n+1}{2}, \quad \dim \mathfrak{n}_n = \binom{n}{2}. \tag{3.1.1}$$

Note that $\mathfrak{d}_n$ and $\mathfrak{n}_n$ are subalgebras of $\mathfrak{b}_n$. Since diagonal matrices commute with each other we have that $\mathfrak{d}_n$ is abelian, whereas easy examples show that $\mathfrak{b}_n$ is non-abelian for $n \ge 2$ and $\mathfrak{n}_n$ is non-abelian for $n \ge 3$. (The smaller cases are trivially abelian, because $\mathfrak{n}_1 = \{0\}$ while $\mathfrak{b}_1$ and $\mathfrak{n}_2$ are one-dimensional.) ■

Example 3.1.4. Let V be a vector space. The linear transformations of V form an associative algebra, which when regarded as a Lie algebra (with commutator bracket) is called $\mathfrak{gl}(V)$. If V has a bilinear multiplication map $V \times V \to V$, there is an important subspace $\mathrm{Der}(V)$ of $\mathfrak{gl}(V)$ consisting of *derivations*, i.e. linear transformations $\varphi : V \to V$ such that

$$\varphi(ab) = \varphi(a)b + a\varphi(b) \quad \text{for all } a, b \in V. \tag{3.1.2}$$

The product (i.e. composition) of two derivations φ and ψ is usually not a derivation, but the commutator $[\varphi, \psi]$ is a derivation, because

$$\begin{aligned} [\varphi,\psi](ab) &= \varphi\psi(ab) - \psi\varphi(ab) = \varphi(\psi(a)b + a\psi(b)) - \psi(\varphi(a)b + a\varphi(b)) \\ &= \Big(\varphi\psi(a)b + \psi(a)\varphi(b) + \varphi(a)\psi(b) + a\,\varphi\psi(b)\Big) \\ &\quad - \Big(\psi\varphi(a)b + \varphi(a)\psi(b) + \psi(a)\varphi(b) + a\,\psi\varphi(b)\Big) \\ &= [\varphi,\psi](a)b + a\,[\varphi,\psi](b). \end{aligned}$$

Hence $\mathrm{Der}(V)$ is a Lie subalgebra of $\mathfrak{gl}(V)$. ■

Example 3.1.4 can be applied to the case where V is a Lie algebra $\mathfrak{g}$ and the multiplication (which we have not assumed to be associative) is the Lie bracket. This results in:

Definition 3.1.5. For any Lie algebra $\mathfrak{g}$, $\mathrm{Der}(\mathfrak{g})$ is the Lie subalgebra of $\mathfrak{gl}(\mathfrak{g})$ consisting of all the derivations of $\mathfrak{g}$, i.e. the linear transformations $\varphi : \mathfrak{g} \to \mathfrak{g}$ satisfying

$$\varphi([x,y]) = [\varphi(x), y] + [x, \varphi(y)] \quad \text{for all } x, y \in \mathfrak{g}. \tag{3.1.3}$$

As with the condition defining a homomorphism, the condition defining a derivation need only be checked in the case where x and y are unequal elements of some chosen basis of $\mathfrak{g}$.

Example 3.1.6. Linear transformations of $\mathfrak{l}_2$ can be identified with 2×2 matrices, using the standard basis s, t of $\mathfrak{l}_2$. The condition defining a derivation of $\mathfrak{l}_2$ reduces to the special case $x = s$, $y = t$. Thus the derivations correspond to those matrices $\left(\begin{smallmatrix} a & b \\ c & d \end{smallmatrix}\right)$ such that

$$bs + dt = [as + ct, t] + [s, bs + dt] = at + dt, \quad \text{i.e. } a = b = 0.$$

So $\mathrm{Der}(\mathfrak{l}_2)$ is isomorphic to the linear Lie algebra $\left(\begin{smallmatrix} 0 & 0 \\ * & * \end{smallmatrix}\right)$, which in turn is isomorphic to $\mathfrak{l}_2$ by Proposition 2.2.7. ■

Many other examples of subalgebras come from the following idea.

Definition 3.1.7. If S is a subset of a Lie algebra $\mathfrak{g}$, the *subalgebra generated by* S is the smallest subalgebra containing S.

More formally (and this is why 'smallest' makes sense), it is the intersection of all subalgebras of $\mathfrak{g}$ containing S: this intersection is clearly itself a subalgebra of $\mathfrak{g}$ containing S. Less formally, it consists of everything you can get by starting with S and taking successive linear combinations and Lie brackets.

Example 3.1.8. A trivial case occurs when S consists of a single element x. The subalgebra generated by x is just $\mathbb{C}x$ (a one-dimensional Lie algebra), since this is clearly closed under the Lie bracket. ■

Example 3.1.9. To give an example where the subalgebra generated by S is larger than the span of S, consider the subalgebra of $\mathfrak{gl}_2$ generated by $e = e_{12}$ and $f = e_{21}$: it also includes $[e, f] = h$, and therefore equals $\mathfrak{sl}_2$. ■

Example 3.1.10. The subalgebra of $\mathfrak{gl}_2$ generated by $e_{11} + e_{12}$ and $e_{21} - e_{22}$ also includes

$$[e_{11} + e_{12}, e_{21} - e_{22}] = e_{11} - e_{12} - e_{21} - e_{22}$$

and

$$[e_{11} + e_{12}, e_{11} - e_{12} - e_{21} - e_{22}] = -e_{11} - 3e_{12} + e_{21} + e_{22}.$$

There is no point in taking further Lie brackets, because the four elements we now have already span $\mathfrak{gl}_2$. Thus $\mathfrak{gl}_2$ is generated by $e_{11} + e_{12}$ and $e_{21} - e_{22}$. There is nothing so special about these elements: if you choose two sufficiently 'random' elements of $\mathfrak{gl}_2$, they will generate the whole Lie algebra (see Exercise 3.4.1). ■

Example 3.1.11. Let us determine the subalgebra of $\mathfrak{gl}_2$ generated by $x = \left(\begin{smallmatrix} 2 & -1 \\ 0 & 1 \end{smallmatrix}\right)$ and $y = \left(\begin{smallmatrix} 1 & 0 \\ 2 & -1 \end{smallmatrix}\right)$. Direct calculation gives

$$[x, y] = \begin{pmatrix} -2 & 2 \\ -2 & 2 \end{pmatrix} = -2x - y + \begin{pmatrix} 3 & 0 \\ 0 & 3 \end{pmatrix}.$$

Hence the subalgebra generated by x and y contains the scalar matrices. The same calculation also shows that $\mathbb{C}\{x, y, 1_2\}$ is closed under the taking of commutators, so this is the subalgebra in question. In the light of the previous example, there must be something special about the pair x, y: in fact, they have a common eigenvector $\left(\begin{smallmatrix} 1 \\ 1 \end{smallmatrix}\right)$, and the subalgebra $\mathbb{C}\{x, y, 1_2\}$ can be described as the set of all matrices having $\left(\begin{smallmatrix} 1 \\ 1 \end{smallmatrix}\right)$ as an eigenvector. Since $\mathfrak{b}_2$ consists of all matrices having $\left(\begin{smallmatrix} 1 \\ 0 \end{smallmatrix}\right)$ as an eigenvector, we have $\mathbb{C}\{x, y, 1_2\} = g\mathfrak{b}_2 g^{-1}$ for any $g \in GL_2$ such that $g\left(\begin{smallmatrix} 1 \\ 0 \end{smallmatrix}\right) = \left(\begin{smallmatrix} 1 \\ 1 \end{smallmatrix}\right)$. Since conjugation by g preserves the Lie bracket, as seen in Exercise 2.3.7, we can conclude that $\mathbb{C}\{x, y, 1_2\}$ is isomorphic to $\mathfrak{b}_2$. ■

Example 3.1.12. The subalgebra of $\mathfrak{gl}_n$ generated by $e_{12}, e_{23}, \ldots, e_{n-1,n}$ is exactly $\mathfrak{n}_n$. To show this, it suffices to show that each element e_{ij}, where $1 \leq i < j \leq n$, lies in the subalgebra; this follows from the easy calculation

$$[\cdots[[e_{i,i+1}, e_{i+1,i+2}], e_{i+2,i+3}], \ldots, e_{j-1,j}] = e_{ij}. \tag{3.1.4}$$

■

3.2 Ideals

The most important subalgebras of a Lie algebra are its ideals, which we can think of as analogous to the normal subgroups of a group.

Definition 3.2.1. An *ideal* of a Lie algebra $\mathfrak{g}$ is a subalgebra $\mathfrak{h}$ such that

$$x \in \mathfrak{g},\ y \in \mathfrak{h} \quad \Rightarrow \quad [x, y] \in \mathfrak{h}.$$

Here 'subalgebra' could be replaced by 'subspace', since the condition implies that $\mathfrak{h}$ is closed under Lie bracket. You might think that this definition gives a 'left' ideal, and that there would be a corresponding notion of 'right' ideal, where the condition is

$$x \in \mathfrak{g},\ y \in \mathfrak{h} \quad \Rightarrow \quad [y, x] \in \mathfrak{h}.$$

But since $[y, x] = -[x, y]$, these conditions are the same. So all ideals are 'two-sided'.

As with ideals of rings and normal subgroups of groups, it is important to be aware that an ideal of a subalgebra is not necessarily an ideal of the whole Lie algebra. However, another implication obviously does hold: any ideal of $\mathfrak{g}$ that is contained in a subalgebra $\mathfrak{h}$ is an ideal of $\mathfrak{h}$ also.

Example 3.2.2. Trivially, $\{0\}$ and $\mathfrak{g}$ are always ideals of $\mathfrak{g}$. By a *nontrivial ideal* we mean an ideal other than these. ■

Example 3.2.3. Any one-dimensional subspace $\mathbb{C}y$ of $\mathfrak{g}$ is a subalgebra, but it is not usually an ideal: for $\mathbb{C}y$ to be an ideal, one would need the extra property that $[x, y] \in \mathbb{C}y$ for all $x \in \mathfrak{g}$. For instance, the only one-dimensional ideal of $\mathfrak{l}_2$ is $\mathbb{C}t$. ■

Definition 3.2.4. The *derived algebra* $\mathcal{D}\mathfrak{g}$ of a Lie algebra $\mathfrak{g}$ is the span of all Lie brackets $[x, y]$, for $x, y \in \mathfrak{g}$. The *centre* of a Lie algebra $\mathfrak{g}$ is

$$Z(\mathfrak{g}) = \{x \in \mathfrak{g} \mid [x, y] = 0 \text{ for all } y \in \mathfrak{g}\}.$$

Proposition 3.2.5. *For any Lie algebra $\mathfrak{g}$, the following are ideals of $\mathfrak{g}$:*

- *$\mathcal{D}\mathfrak{g}$, and any subspace of $\mathfrak{g}$ that contains $\mathcal{D}\mathfrak{g}$;*
- *$Z(\mathfrak{g})$, and any subspace of $Z(\mathfrak{g})$.*

Proof. These are easy consequences of the definitions. □

Example 3.2.6. To say that $\mathfrak{g}$ is abelian is equivalent to saying that $\mathcal{D}\mathfrak{g} = \{0\}$ and is also equivalent to saying that $Z(\mathfrak{g}) = \mathfrak{g}$. If this is the case then all subspaces of $\mathfrak{g}$ are ideals of $\mathfrak{g}$. ■

Proposition 3.2.7. *If $\mathfrak{h}_1$ and $\mathfrak{h}_2$ are ideals of $\mathfrak{g}$, then so are $\mathfrak{h}_1 + \mathfrak{h}_2$, $\mathfrak{h}_1 \cap \mathfrak{h}_2$, and*

$$[\mathfrak{h}_1, \mathfrak{h}_2] = \mathbb{C}\{[y, z] \mid y \in \mathfrak{h}_1, z \in \mathfrak{h}_2\}.$$

Moreover, $[\mathfrak{h}_1, \mathfrak{h}_2]$ is contained in $\mathfrak{h}_1 \cap \mathfrak{h}_2$.

Proof. The only statement that is not immediate from the definitions is that $[\mathfrak{h}_1, \mathfrak{h}_2]$ is an ideal of $\mathfrak{g}$. This follows from the Jacobi identity: if $x \in \mathfrak{g}$, $y \in \mathfrak{h}_1$, $z \in \mathfrak{h}_2$ then

$$[x, [y, z]] = [[x, y], z] + [y, [x, z]] \in [\mathfrak{h}_1, \mathfrak{h}_2],$$

and hence $[x, w] \in [\mathfrak{h}_1, \mathfrak{h}_2]$ for all $w \in [\mathfrak{h}_1, \mathfrak{h}_2]$, as required. □

Remark 3.2.8. Despite what might be suggested by a superficial interpretation of the notation, $[\mathfrak{h}_1, \mathfrak{h}_2]$ consists of all linear combinations of Lie brackets $[y, z]$ with $y \in \mathfrak{h}_1$ and $z \in \mathfrak{h}_2$, not just these Lie brackets themselves. Nevertheless, in proofs involving an ideal of the form $[\mathfrak{h}_1, \mathfrak{h}_2]$, often only a single Lie bracket $[y, z]$ is considered; this is valid in situations where the general statement can be deduced from that case (as in the proof of Proposition 3.2.7). The same comment applies to $\mathcal{D}\mathfrak{g}$, which is merely an alternative notation for $[\mathfrak{g}, \mathfrak{g}]$.

Example 3.2.9. It is easy to find examples showing that, for $n \geq 2$, the upper-triangular subalgebra $\mathfrak{b}_n$ and the strictly upper-triangular subalgebra $\mathfrak{n}_n$ are not ideals of $\mathfrak{gl}_n$, nor is $\mathfrak{d}_n$ an ideal of $\mathfrak{b}_n$. However, $\mathfrak{n}_n$ is an ideal of $\mathfrak{b}_n$: more precisely, we claim that $\mathfrak{n}_n = \mathcal{D}\mathfrak{b}_n$. The fact that $e_{ij} = [e_{ii}, e_{ij}]$ for $1 \leq i < j \leq n$ proves that $\mathfrak{n}_n \subseteq \mathcal{D}\mathfrak{b}_n$. To show the reverse inclusion, we must show that, for any two upper-triangular $n \times n$ matrices x and y, $[x, y] = xy - yx$ is strictly upper-triangular. This comes from the familiar fact that the diagonal entries of upper-triangular matrices multiply position by position, so that xy and yx have the same diagonal; this makes the diagonal entries of $[x, y]$ zero. ■

It is an important result in the theory of associative algebras that Mat_n has no nontrivial two-sided ideals (in the sense of rings). By contrast, the Lie algebra $\mathfrak{gl}_n$ does have nontrivial ideals when $n \geq 2$: two such ideals are determined in the following result.

Proposition 3.2.10. *We have* $\mathcal{D}\mathfrak{gl}_n = \mathfrak{sl}_n$ *and* $Z(\mathfrak{gl}_n) = \mathbb{C}1_n$ *(the scalar matrices).*

Proof. By (2.1.1), we have $\mathcal{D}\mathfrak{gl}_n \subseteq \mathfrak{sl}_n$. To prove the reverse inclusion (i.e. that every trace-zero matrix is a linear combination of commutators), we use the basis of $\mathfrak{sl}_n$ consisting of

$$e_{ij} \text{ for } 1 \leq i \neq j \leq n \quad \text{and} \quad e_{ii} - e_{i+1,i+1} \text{ for } 1 \leq i \leq n-1.$$

It is easy to see that all these basis elements are commutators: specifically, $e_{ij} = [e_{ii}, e_{ij}]$ if $i \neq j$ and $e_{ii} - e_{i+1,i+1} = [e_{i,i+1}, e_{i+1,i}]$.

Since scalar matrices commute with any matrix, we certainly have $\mathbb{C}1_n \subseteq Z(\mathfrak{gl}_n)$. The reverse inclusion (i.e. that any $n \times n$ matrix that commutes with all $n \times n$ matrices is a scalar matrix) can be proved by commutator calculations with the standard basis (see Exercise 3.4.4); but the following more conceptual argument is neater. Suppose that $x \in Z(\mathfrak{gl}_n)$ and $y \in \mathfrak{gl}_n$. Let $W \subseteq \mathbb{C}^n$ be any nonzero eigenspace of x, say for the eigenvalue $a \in \mathbb{C}$. Since y commutes with x, it must preserve W; the proof, the reader may remember, is that for any $w \in W$ we have

$$xyw = yxw = y(aw) = ayw,$$

which shows that $yw \in W$. Hence W is preserved by the action of every matrix. But this forces W to be the whole of $\mathbb{C}^n$, since there are matrices that take any nonzero element of $\mathbb{C}^n$ to any other nonzero element of $\mathbb{C}^n$. This means that x acts as scalar multiplication by a on the whole of $\mathbb{C}^n$, i.e. x is the scalar matrix $a1_n$. □

Theorem 3.2.11. *Every ideal of* $\mathfrak{gl}_n$ *is either* $\{0\}$, $\mathbb{C}1_n$, $\mathfrak{sl}_n$, *or* $\mathfrak{gl}_n$.

Proof. Let $\mathfrak{h}$ be an ideal of $\mathfrak{gl}_n$. Since $\mathbb{C}1_n$ is one-dimensional and $\mathfrak{sl}_n$ has codimension 1 in $\mathfrak{gl}_n$, it suffices to show that either $\mathfrak{h} \subseteq \mathbb{C}1_n$ or $\mathfrak{h} \supseteq \mathfrak{sl}_n$.

We first consider the case $\mathfrak{h} \subseteq \mathfrak{d}_n$. If it is not true that $\mathfrak{h} \subseteq \mathbb{C}1_n$ then $\mathfrak{h}$ contains a diagonal matrix $\sum_k a_{kk}e_{kk}$ where $a_{ii} \neq a_{jj}$ for some $i \neq j$. Since $\mathfrak{h}$ is an ideal of $\mathfrak{gl}_n$, it must also contain $[\sum_k a_{kk}e_{kk}, e_{ij}] = (a_{ii} - a_{jj})e_{ij}$ and hence e_{ij}, which contradicts the assumption that $\mathfrak{h} \subseteq \mathfrak{d}_n$. So in this case we must have $\mathfrak{h} \subseteq \mathbb{C}1_n$.

Now suppose that $\mathfrak{h} \not\subseteq \mathfrak{d}_n$. Hence there is an element $x = \sum_{k,l} a_{kl}e_{kl} \in \mathfrak{h}$ such that $a_{ij} \neq 0$ for some particular pair (i, j) with $i \neq j$. We deduce successively that $\mathfrak{h}$ contains the following elements:

$$\begin{aligned} [e_{ii}, x] &= \sum_l a_{il}e_{il} - \sum_k a_{ki}e_{ki}, \\ [[e_{ii}, x], e_{jj}] &= a_{ij}e_{ij} + a_{ji}e_{ji}, \\ [e_{ii}, [[e_{ii}, x], e_{jj}]] &= a_{ij}e_{ij} - a_{ji}e_{ji}. \end{aligned}$$

So $\mathfrak{h}$ contains $2a_{ij}e_{ij}$ and hence e_{ij}. Then it must also contain $e_{ik} = [e_{ij}, e_{jk}]$ for all $k \neq i, j$, so it contains e_{il} for all $l \neq i$. Hence it must also contain $e_{kl} = [e_{ki}, e_{il}]$ for all (k, l) where $k \neq l$ and $l \neq i$. Also, $\mathfrak{h}$ contains $-2e_{ki} = [[e_{ik}, e_{ki}], e_{ki}]$ for all $k \neq i$, so in fact $\mathfrak{h}$ contains all the basis elements e_{kl} where $k \neq l$. Therefore it also contains $e_{kk} - e_{k+1,k+1} = [e_{k,k+1}, e_{k+1,k}]$ for all $1 \leq k \leq n - 1$, so it contains a whole basis of $\mathfrak{sl}_n$; thus $\mathfrak{h} \supseteq \mathfrak{sl}_n$. The proof is finished. □

We will often use the direct sum decomposition

$$\mathfrak{gl}_n = \mathbb{C}1_n \oplus \mathfrak{sl}_n, \tag{3.2.1}$$

which says in words that an $n \times n$ matrix can be written uniquely as the sum of a scalar matrix and a matrix with trace zero. The proof is simply that $\mathbb{C}1_n \cap \mathfrak{sl}_n = \{0\}$ because the trace of $a1_n$ is an, and $\dim \mathbb{C}1_n + \dim \mathfrak{sl}_n = 1 + (n^2 - 1) = \dim \mathfrak{gl}_n$.

3.3 Quotients and simple Lie algebras

As is the case for two-sided ideals of rings and normal subgroups of groups, one may form the quotient of a Lie algebra by an ideal.

Proposition 3.3.1. *If* $\mathfrak{h}$ *is an ideal of* $\mathfrak{g}$ *then the quotient space* $\mathfrak{g}/\mathfrak{h}$ *has a Lie algebra structure with bracket*

$$[x + \mathfrak{h}, y + \mathfrak{h}] = [x, y] + \mathfrak{h}.$$

Proof. To show that this bracket is well defined, let $x', y' \in \mathfrak{h}$; then

$$[x + x', y + y'] = [x, y] + [x, y'] + [x', y] + [x', y'] \in [x, y] + \mathfrak{h},$$

as required. The three Lie algebra axioms follow trivially from the corresponding properties for $\mathfrak{g}$. □

This definition makes the canonical map $\mathfrak{g} \to \mathfrak{g}/\mathfrak{h}$ a homomorphism.

We have the usual fundamental homomorphism theorem:

Proposition 3.3.2. *If $\varphi : \mathfrak{g} \to \mathfrak{g}'$ is a homomorphism of Lie algebras then*

$$\ker(\varphi) = \varphi^{-1}(\{0\}) = \{x \in \mathfrak{g} \mid \varphi(x) = 0\}$$

is an ideal of $\mathfrak{g}$, and

$$\operatorname{im}(\varphi) = \varphi(\mathfrak{g}) = \{\varphi(x) \mid x \in \mathfrak{g}\}$$

is a subalgebra of $\mathfrak{g}'$. Moreover, the induced map

$$\psi : \mathfrak{g}/\ker(\varphi) \to \operatorname{im}(\varphi) : x + \ker(\varphi) \mapsto \varphi(x)$$

is an isomorphism of Lie algebras.

Proof. If $x \in \mathfrak{g}$ and $y \in \ker(\varphi)$, we have

$$\varphi([x, y]) = [\varphi(x), \varphi(y)] = [\varphi(x), 0] = 0;$$

so $[x, y] \in \ker(\varphi)$, whence $\ker(\varphi)$ is an ideal. That $\varphi(\mathfrak{g})$ is a subalgebra of $\mathfrak{g}'$ is even more obvious. The map ψ is bijective, by the corresponding result for vector spaces, and is a Lie algebra homomorphism because

$$\begin{aligned}\psi([x + \ker(\varphi), y + \ker(\varphi)]) &= \psi([x, y] + \ker(\varphi)) = \varphi([x, y]) \\ &= [\varphi(x), \varphi(y)] = [\psi(x + \ker(\varphi)), \psi(y + \ker(\varphi))],\end{aligned}$$

where the first step uses the definition in Proposition 3.3.1 of a quotient Lie algebra. □

Proposition 3.3.3. *If $\varphi : \mathfrak{g} \to \mathfrak{g}'$ is a surjective homomorphism of Lie algebras then there is a bijection between the subalgebras of $\mathfrak{g}$ containing $\ker(\varphi)$ and the subalgebras of $\mathfrak{g}'$. In one direction, this bijection sends $\mathfrak{h}$ to $\varphi(\mathfrak{h})$; in the other direction, it sends $\mathfrak{h}'$ to $\varphi^{-1}(\mathfrak{h}')$. Under this bijection the ideals of $\mathfrak{g}$ containing $\ker(\varphi)$ correspond to the ideals of $\mathfrak{g}'$.*

Proof. It is immediate from the relevant definitions that if $\mathfrak{h}$ is a subalgebra (respectively, ideal) of $\mathfrak{g}$ then $\varphi(\mathfrak{h})$ is a subalgebra (respectively, ideal) of $\mathfrak{g}'$; and if $\mathfrak{h}'$ is a subalgebra (respectively, ideal) of $\mathfrak{g}'$ then $\varphi^{-1}(\mathfrak{h}')$ is a subalgebra (respectively, ideal) of $\mathfrak{g}$. Moreover, if $\mathfrak{h}$ contains $\ker(\varphi)$ then $\varphi^{-1}(\varphi(\mathfrak{h})) = \mathfrak{h}$, and, for any $\mathfrak{h}'$, $\varphi(\varphi^{-1}(\mathfrak{h}')) = \mathfrak{h}'$ because φ is surjective. The result follows. □

In particular, the subalgebras (respectively, ideals) of $\mathfrak{g}/\mathfrak{h}$ are in bijection with the subalgebras (respectively, ideals) of $\mathfrak{g}$ containing $\mathfrak{h}$.

Proposition 3.3.4. *If $\varphi : \mathfrak{g} \to \mathfrak{g}'$ is a Lie algebra homomorphism then $\varphi(\mathcal{D}\mathfrak{g}) \subseteq \mathcal{D}\mathfrak{g}'$. If φ is surjective then we have equality: $\varphi(\mathcal{D}\mathfrak{g}) = \mathcal{D}\mathfrak{g}'$.*

Proof. This all follows easily from the equation $\varphi([x,y]) = [\varphi(x), \varphi(y)]$ for $x, y \in \mathfrak{g}$. (To reiterate the warning in Remark 3.2.8, $\mathcal{D}\mathfrak{g}$ does not consist only of such $[x,y]$ but is spanned by such $[x,y]$). □

Corollary 3.3.5. *The quotient $\mathfrak{g}/\mathfrak{h}$ is abelian if and only if $\mathfrak{h} \supseteq \mathcal{D}\mathfrak{g}$.*

Proof. We can rewrite the condition for the quotient to be abelian as $\mathcal{D}(\mathfrak{g}/\mathfrak{h}) = \{0\}$, which by Proposition 3.3.4 is equivalent to saying that $\mathcal{D}\mathfrak{g}$ is in the kernel of the projection $\mathfrak{g} \to \mathfrak{g}/\mathfrak{h}$. □

For this reason, $\mathfrak{g}/\mathcal{D}\mathfrak{g}$ is called the *maximal abelian quotient* of $\mathfrak{g}$.

Example 3.3.6. Since $\mathcal{D}\,\mathfrak{l}_2 = \mathbb{C}t$, the maximal abelian quotient of $\mathfrak{l}_2$ is the one-dimensional Lie algebra $\mathfrak{l}_2/\mathbb{C}t$. ■

Example 3.3.7. We saw above that $\mathcal{D}\mathfrak{b}_n = \mathfrak{n}_n$. Hence the maximal abelian quotient of $\mathfrak{b}_n$ is $\mathfrak{b}_n/\mathfrak{n}_n$, which is n-dimensional. In fact, by the fundamental homomorphism theorem, the homomorphism $\mathfrak{b}_n \to \mathfrak{d}_n$ that annuls all above-diagonal entries induces an isomorphism $\mathfrak{b}_n/\mathfrak{n}_n \xrightarrow{\sim} \mathfrak{d}_n$. ■

If $\mathfrak{g}$ has a nontrivial ideal $\mathfrak{g}_1$ then $\mathfrak{g}$ can be regarded as being built from the smaller-dimensional Lie algebras $\mathfrak{g}_1$ and $\mathfrak{g}/\mathfrak{g}_1$. The basic 'building-blocks', from this viewpoint, are the simple Lie algebras, analogues of simple groups or rings:

Definition 3.3.8. A Lie algebra $\mathfrak{g}$ is *simple* if its only ideals are $\{0\}$ and $\mathfrak{g}$ and its dimension is more than 1.

It is conventional to exclude the one-dimensional Lie algebra $\mathbb{C}$ in this way. One major difference between $\mathbb{C}$ and all the simple Lie algebras is that $\mathbb{C}$ is abelian whereas simple Lie algebras are not:

Proposition 3.3.9. *If $\mathfrak{g}$ is simple then $\mathcal{D}\mathfrak{g} = \mathfrak{g}$ and $Z(\mathfrak{g}) = \{0\}$.*

Proof. Since $\mathfrak{g}$ has no nontrivial ideals, $\mathcal{D}\mathfrak{g}$ must be either $\{0\}$ or $\mathfrak{g}$. But $\mathcal{D}\mathfrak{g} = \{0\}$ would mean that $\mathfrak{g}$ is abelian, so that every subspace would be an ideal and that would force $\mathfrak{g}$ to be one-dimensional, contrary to our assumption. Hence $\mathcal{D}\mathfrak{g} = \mathfrak{g}$. The argument for $Z(\mathfrak{g})$ is similar. □

We saw in Theorem 3.2.11 that $\mathfrak{gl}_n$ is never simple. Instead we have:

Theorem 3.3.10. *For all $n \geq 2$, $\mathfrak{sl}_n$ is simple.*

Proof. Suppose that $\mathfrak{h}$ is an ideal of $\mathfrak{sl}_n$. Then $[x, \mathfrak{h}] \subseteq \mathfrak{h}$ for all $x \in \mathfrak{sl}_n$. Since $[a1_n, \mathfrak{h}] = \{0\}$ for all $a \in \mathbb{C}$, this implies that $[x + a1_n, \mathfrak{h}] \subseteq \mathfrak{h}$ for all $x \in \mathfrak{sl}_n$ and $a \in \mathbb{C}$. As was observed in (3.2.1), every element of $\mathfrak{gl}_n$ is such a sum $x + a1_n$ so $\mathfrak{h}$ is actually an ideal of $\mathfrak{gl}_n$. By Theorem 3.2.11, the only ideals of $\mathfrak{gl}_n$ contained in $\mathfrak{sl}_n$ are $\{0\}$ and $\mathfrak{sl}_n$, which finishes the proof. □

From our description of two-dimensional Lie algebras we can see that none of them is simple, so the three-dimensional Lie algebra $\mathfrak{sl}_2$ is as small as a simple Lie algebra can be. This is part of the reason for its great importance in the theory.

Remark 3.3.11. A great theorem of the twentieth century was the Cartan–Killing classification of simple Lie algebras up to isomorphism. Its full statement, let alone its proof, is beyond the scope of this book; but we will say more about it in Section 8.1. We merely mention here that $\mathfrak{so}_n$ is simple for all $n \geq 5$. (Exercise 2.3.5 shows that $\mathfrak{so}_3$ is isomorphic to $\mathfrak{sl}_2$, so it is also simple; we will see in Exercise 6.5.7 that $\mathfrak{so}_4$ is not simple.)

Changing to an opposite train of thought, we need to consider how to combine two Lie algebras into a larger one. The easiest way is the following.

Definition 3.3.12. If $\mathfrak{g}_1$ and $\mathfrak{g}_2$ are Lie algebras, the *direct product* $\mathfrak{g}_1 \times \mathfrak{g}_2$ is the Lie algebra that, as a set, is the Cartesian product, with componentwise operations:

$$\begin{aligned} a(x_1, x_2) + b(y_1, y_2) &= (ax_1 + by_1, ax_2 + by_2), \\ [(x_1, x_2), (y_1, y_2)] &= ([x_1, y_1], [x_2, y_2]), \end{aligned}$$

for all $a, b \in \mathbb{C}$, $x_1, y_1 \in \mathfrak{g}_1$, $x_2, y_2 \in \mathfrak{g}_2$.

The proof that this is a Lie algebra is routine. Note that, as a vector space, $\mathfrak{g}_1 \times \mathfrak{g}_2$ is what would normally be called the external direct sum $\mathfrak{g}_1 \oplus \mathfrak{g}_2$. Because the two factors $\mathfrak{g}_1$ and $\mathfrak{g}_2$ do not 'interfere' with each other's Lie brackets, for many purposes they can be considered independently. For instance, it is easy to see that

$$\mathcal{D}(\mathfrak{g}_1 \times \mathfrak{g}_2) = \mathcal{D}\mathfrak{g}_1 \times \mathcal{D}\mathfrak{g}_2 \quad \text{and} \quad Z(\mathfrak{g}_1 \times \mathfrak{g}_2) = Z(\mathfrak{g}_1) \times Z(\mathfrak{g}_2). \tag{3.3.1}$$

Clearly $\mathfrak{g}_1 \times \{0\}$ is an ideal of $\mathfrak{g}_1 \times \mathfrak{g}_2$ isomorphic to $\mathfrak{g}_1$ via the obvious map, and $\{0\} \times \mathfrak{g}_2$ is an ideal of $\mathfrak{g}_1 \times \mathfrak{g}_2$ isomorphic to $\mathfrak{g}_2$ via the obvious map. Moreover the projection maps $\mathfrak{g}_1 \times \mathfrak{g}_2 \to \mathfrak{g}_1$ and $\mathfrak{g}_1 \times \mathfrak{g}_2 \to \mathfrak{g}_2$ are homomorphisms, and applying the fundamental homomorphism theorem to them gives isomorphisms $(\mathfrak{g}_1 \times \mathfrak{g}_2)/(\{0\} \times \mathfrak{g}_2) \cong \mathfrak{g}_1$ and $(\mathfrak{g}_1 \times \mathfrak{g}_2)/(\mathfrak{g}_1 \times \{0\}) \cong \mathfrak{g}_2$. Unless it creates confusion (for instance, if $\mathfrak{g}_1$ and $\mathfrak{g}_2$ are the same), one commonly identifies $\mathfrak{g}_1 \times \{0\}$

with $\mathfrak{g}_1$ and $\{0\} \times \mathfrak{g}_2$ with $\mathfrak{g}_2$. Additional justification for this abuse of notation is provided by:

Proposition 3.3.13. *Suppose that $\mathfrak{g}$ is a Lie algebra with a direct sum decomposition $\mathfrak{g} = \mathfrak{g}_1 \oplus \mathfrak{g}_2$, where $\mathfrak{g}_1$ and $\mathfrak{g}_2$ are ideals of $\mathfrak{g}$. Then the map $\mathfrak{g}_1 \times \mathfrak{g}_2 \to \mathfrak{g} : (x_1, x_2) \mapsto x_1 + x_2$ is an isomorphism of Lie algebras.*

Proof. The only thing that requires proof is that this map respects the Lie bracket, i.e. that $[x_1 + x_2, y_1 + y_2] = [x_1, y_1] + [x_2, y_2]$ for all $x_1, y_1 \in \mathfrak{g}_1$, $x_2, y_2 \in \mathfrak{g}_2$. This holds because $[\mathfrak{g}_1, \mathfrak{g}_2] \subseteq \mathfrak{g}_1 \cap \mathfrak{g}_2 = \{0\}$. □

Example 3.3.14. Applying Proposition 3.3.13 to the direct sum in (3.2.1) gives the isomorphism $\mathfrak{gl}_n \cong \mathbb{C} \times \mathfrak{sl}_n$. ■

Example 3.3.15. The direct sum decomposition $\mathfrak{b}_n = \mathfrak{d}_n \oplus \mathfrak{n}_n$ is not of the same kind, because $\mathfrak{d}_n$ is not an ideal of $\mathfrak{b}_n$. So we cannot conclude that $\mathfrak{b}_n \cong \mathfrak{d}_n \times \mathfrak{n}_n$; in fact $\mathfrak{b}_n$ is definitely not isomorphic to $\mathfrak{d}_n \times \mathfrak{n}_n$, because $\mathcal{D}\mathfrak{b}_n = \mathfrak{n}_n$ has a larger dimension than $\mathcal{D}(\mathfrak{d}_n \times \mathfrak{n}_n) = \{0\} \times \mathcal{D}\mathfrak{n}_n$. ■

As the last example shows, we cannot take the isomorphism $\mathfrak{g}/\mathfrak{g}_1 \cong \mathfrak{g}_2$ and deduce the isomorphism $\mathfrak{g} \cong \mathfrak{g}_1 \times \mathfrak{g}_2$. So direct products are far from the whole story about how an ideal $\mathfrak{g}_1$ and a quotient $\mathfrak{g}/\mathfrak{g}_1$ can combine to form $\mathfrak{g}$.

3.4 Exercises

Exercise 3.4.1. Let $x = ae_{11} + be_{12} + ce_{21} + de_{22} \in \mathfrak{gl}_2$, where $a, b, c, d \in \mathbb{C}$.

(i) Show that if $bc \neq 0$ then x and e_{11} generate the whole of $\mathfrak{gl}_2$.
(i) If $bc = 0$, what possibilities are there for the subalgebra of $\mathfrak{gl}_2$ generated by x and e_{11}?

Exercise 3.4.2. Let A be an associative $\mathbb{C}$-algebra. We can define two different Lie algebras of derivations of A: $\mathrm{Der}(A)$, which consists of derivations of A relative to the associative multiplication; and $\mathrm{Der}_{\mathrm{Lie}}(A)$, which consists of derivations of A relative to the commutator bracket (in other words, derivations of the Lie algebra A).

(i) Show that $\mathrm{Der}(A)$ is a subalgebra of $\mathrm{Der}_{\mathrm{Lie}}(A)$.
(ii) Give an example for A for which $\mathrm{Der}(A)$ is not an ideal of $\mathrm{Der}_{\mathrm{Lie}}(A)$.

Exercise 3.4.3. Suppose that $\mathfrak{h}$ is an ideal of $\mathfrak{g}$ with $\dim \mathfrak{g} - \dim \mathfrak{h} = k$.

(i) Show that if $k = 1$ or $k = 2$ then $\mathcal{D}\mathfrak{g} \neq \mathfrak{g}$.
(ii) Show that if $\mathfrak{h} = Z(\mathfrak{g})$ then $\dim \mathcal{D}\mathfrak{g} \leq \binom{k}{2}$.
(iii) Hence show that it is impossible to have $\dim Z(\mathfrak{g}) = \dim \mathfrak{g} - 1$.

Exercise 3.4.4. For any subset S of a Lie algebra $\mathfrak{g}$, define its *centralizer* $Z_{\mathfrak{g}}(S)$ as $\{x \in \mathfrak{g} \mid [x, y] = 0, \text{ for all } y \in S\}$.

(i) Show that $Z_{\mathfrak{g}}(S)$ is a subalgebra of $\mathfrak{g}$.
(ii) Show that if $\mathfrak{h}$ is an ideal of $\mathfrak{g}$, $Z_{\mathfrak{g}}(\mathfrak{h})$ is also an ideal of $\mathfrak{g}$.
(iii) Determine $Z_{\mathfrak{gl}_n}(\mathfrak{d}_n)$.
(iv) Using the previous part, give another proof that $Z(\mathfrak{gl}_n) = \mathbb{C}1_n$.

Exercise 3.4.5. For any subalgebra $\mathfrak{h}$ of a Lie algebra $\mathfrak{g}$, define its *normalizer* $N_{\mathfrak{g}}(\mathfrak{h})$ as $\{x \in \mathfrak{g} \mid [x, \mathfrak{h}] \subseteq \mathfrak{h}\}$.

(i) Show that $N_{\mathfrak{g}}(\mathfrak{h})$ is a subalgebra of $\mathfrak{g}$ and $\mathfrak{h}$ is an ideal of $N_{\mathfrak{g}}(\mathfrak{h})$.
(ii) Show that $Z_{\mathfrak{g}}(\mathfrak{h})$ is an ideal of $N_{\mathfrak{g}}(\mathfrak{h})$.
(iii) Determine $N_{\mathfrak{gl}_n}(\mathfrak{n}_n)$.

Exercise 3.4.6. Suppose that $\dim \mathfrak{g} = 3$ and $\dim \mathcal{D}\mathfrak{g} = 1$.

(i) Show that if $\mathcal{D}\mathfrak{g} \subseteq Z(\mathfrak{g})$ then $\mathfrak{g} \cong \mathfrak{n}_3$.
(ii) Show that if $\mathcal{D}\mathfrak{g} \not\subseteq Z(\mathfrak{g})$ then $\mathfrak{g} \cong \mathbb{C} \times \mathfrak{l}_2$.

(*Hint*: to get started, choose a basis x, y, z of $\mathfrak{g}$ such that $\mathcal{D}\mathfrak{g} = \mathbb{C}z$. Then we must have $[x, y] = az$, $[x, z] = bz$, $[y, z] = cz$ for some $a, b, c \in \mathbb{C}$ that are not all zero.)

Exercise 3.4.7. Suppose that $\dim \mathfrak{g} = 3$ and $\dim \mathcal{D}\mathfrak{g} = 2$. Show that $\mathfrak{g}$ is isomorphic to one of the Lie algebras in Exercise 2.3.8. What is $\dim Z(\mathfrak{g})$?

Exercise 3.4.8. If $\mathfrak{h}$ is an ideal of $\mathfrak{g}$, we say that a subalgebra $\mathfrak{h}'$ of $\mathfrak{g}$ is *complementary* to $\mathfrak{h}$ if $\mathfrak{g} = \mathfrak{h} \oplus \mathfrak{h}'$ (a direct sum of subspaces).

(i) Show that if $\mathfrak{h}'$ is such a complementary subalgebra then $\mathfrak{h}' \cong \mathfrak{g}/\mathfrak{h}$.
(ii) Give an example of $\mathfrak{g}$ and $\mathfrak{h}$ such that no complementary subalgebra exists.
(iii) Show that $\mathfrak{g}$ has a subalgebra complementary to $\mathfrak{h}$ if and only if there is a Lie algebra homomorphism $\varphi : \mathfrak{g}/\mathfrak{h} \to \mathfrak{g}$ such that $\varphi(x + \mathfrak{h}) \in x + \mathfrak{h}$ for all $x \in \mathfrak{g}$.

Exercise 3.4.9. Suppose that $\dim \mathfrak{g} = m \geq 4$, $\dim \mathcal{D}\mathfrak{g} = 1$, and $\mathcal{D}\mathfrak{g} \not\subseteq Z(\mathfrak{g})$. Show that $\mathfrak{g}$ is the direct sum of an $(m-2)$-dimensional abelian ideal and a two-dimensional non-abelian ideal.

CHAPTER 4

Modules over a Lie algebra

As discussed in Chapter 1, one of the most important algebraic problems relating to a Lie algebra $\mathfrak{g}$ is to classify its *representations*. In simple matrix terms, these are the Lie algebra homomorphisms $\mathfrak{g} \to \mathfrak{gl}_n$ for various n. But it is better to be more notationally flexible and to define the representations of $\mathfrak{g}$ to be the Lie algebra homomorphisms $\mathfrak{g} \to \mathfrak{gl}(V)$, where V is a vector space and $\mathfrak{gl}(V)$ denotes the Lie algebra of linear transformations of V; these linear transformations can of course be identified with matrices, once a basis of V is chosen. Having done this, it is just a small shift of viewpoint to regard the vector space V as the primary object (at least notationally) and the representation as an extra structure that it carries; this can then be rephrased in terms reminiscent of modules over a ring.

4.1 Definition of a module

The basic definition is as follows.

Definition 4.1.1. Let $\mathfrak{g}$ be a Lie algebra. A $\mathfrak{g}$-*module* is a vector space V equipped with a bilinear map $\mathfrak{g} \times V \to V : (x, v) \mapsto xv$, called the $\mathfrak{g}$-*action*, which is assumed to satisfy

$$[x, y]v = x(yv) - y(xv) \quad \text{for all } x, y \in \mathfrak{g},\ v \in V.$$

For $x \in \mathfrak{g}$, we write x_V for the linear transformation $V \to V : v \mapsto xv$; this linear transformation is called the *action of x on V*.

The condition defining an action can be rewritten as

$$[x, y]_V = [x_V, y_V] \quad \text{for all } x, y \in \mathfrak{g}, \tag{4.1.1}$$

where the bracket on the left-hand side is the Lie bracket in $\mathfrak{g}$ and the bracket on the right-hand side is the Lie bracket (i.e. commutator) in $\mathfrak{gl}(V)$. So the condition (4.1.1) defining an action is equivalent to saying that the map $\mathfrak{g} \to \mathfrak{gl}(V) : x \mapsto x_V$ is a homomorphism. Such a homomorphism is called a *representation of* $\mathfrak{g}$ *on* V; thus, specifying a $\mathfrak{g}$-action on V is equivalent to specifying a representation of $\mathfrak{g}$ on V.

Remark 4.1.2. Many authors put the notational emphasis on the representation by giving it its own letter, such as $\pi : \mathfrak{g} \to \mathfrak{gl}(V)$. Here we follow Bourbaki [1] in putting the emphasis on the vector space and writing the representation simply as $x \mapsto x_V$ (with a few exceptions mentioned below).

Definition 4.1.3. If V is a $\mathfrak{g}$-module, we write $\mathfrak{g}_V = \{x_V \mid x \in \mathfrak{g}\}$ for the image of the representation, a subalgebra of $\mathfrak{gl}(V)$. We say that V is a *faithful* $\mathfrak{g}$-module if the representation is injective, i.e. it gives an isomorphism $\mathfrak{g} \xrightarrow{\sim} \mathfrak{g}_V$. The term 'faithful' is applied to the action and to the representation also.

Speaking philosophically, the idea here is that the Lie algebra $\mathfrak{g}$ encodes some abstract commutator relations in its Lie bracket and that we are considering the occurrences of these commutator relations among concrete linear transformations or matrices. These linear transformations or matrices 'represent' the elements of the Lie algebra, in that they obey the same bracket relations. They do so 'faithfully' if they do not satisfy any extra relations.

If $\mathfrak{g}$ has a basis $x_1, \dots, x_m$ then to specify a $\mathfrak{g}$-action on a vector space V it suffices to specify the linear transformations $(x_1)_V, \dots, (x_m)_V$ that give the action of the basis elements, because a general element $a_1 x_1 + \cdots + a_m x_m$ must act as $a_1(x_1)_V + \cdots + a_m(x_m)_V$. If a basis of V has also been chosen, these m linear transformations can be specified by their matrices, the 'representing matrices' of the representation. Of course, these linear transformations or matrices cannot be chosen freely: by (4.1.1), they have to satisfy the bracket relations satisfied by the chosen basis elements of $\mathfrak{g}$.

Example 4.1.4. Recall the Lie algebra $\mathfrak{l}_2$ with basis s, t and defining relation $[s, t] = t$. We can rephrase the definition of an $\mathfrak{l}_2$-module as follows: it is a vector space V equipped with two linear transformations s_V and t_V such that $[s_V, t_V] = t_V$. For example, let $V = \mathbb{C}^2$ and define the $\mathfrak{l}_2$-action in such a way that $s_{\mathbb{C}^2}$ is the matrix $\left(\begin{smallmatrix} 1 & 0 \\ 0 & 0 \end{smallmatrix}\right)$ and $t_{\mathbb{C}^2}$ is the matrix $\left(\begin{smallmatrix} 0 & 1 \\ 0 & 0 \end{smallmatrix}\right)$. We have checked the requisite commutation relation already: this representation is the isomorphism of $\mathfrak{l}_2$ with $\left(\begin{smallmatrix} * & * \\ 0 & 0 \end{smallmatrix}\right)$ seen in Example 2.2.3. In particular, $\mathbb{C}^2$ is a faithful $\mathfrak{l}_2$-module. ■

Example 4.1.5. Similarly, an $\mathfrak{sl}_2$-module is a vector space V equipped with three linear transformations e_V, h_V, and f_V, such that

$$[e_V, f_V] = h_V, \quad [h_V, e_V] = 2e_V, \quad [h_V, f_V] = -2f_V.$$

We will classify all $\mathfrak{sl}_2$-modules in the next chapter. ■

Example 4.1.6. If A is an associative algebra, a module over A in the usual sense of rings is also a module over A regarded as a Lie algebra (with commutator bracket), since $(ab)v = a(bv)$ and $(ba)v = b(av)$ together imply $(ab-ba)v = a(bv) - b(av)$. But the converse is usually false. For instance, we will see many $\mathfrak{gl}_n$-modules that are not Mat_n-modules, i.e. the linear transformations representing the matrices in $\mathfrak{gl}_n$ do not multiply as these matrices do although their commutators match up. (We have actually seen such a case already, in Example 1.3.1.) This is a main reason for introducing the separate notation $\mathfrak{gl}_n$ for Mat_n regarded as a Lie algebra. ■

Suppose that V is a $\mathfrak{g}$-module. To minimize the use of parentheses, we will often write an expression such as $x(yv)$, the result of acting on $v \in V$ first by $y \in \mathfrak{g}$ and then by $x \in \mathfrak{g}$, simply as xyv and an expression such as $x(xv)$ simply as x^2v. It is important to bear in mind that these do not (in general) have anything to do with products xy and x^2 taken in some associative algebra to which x and y may happen to belong. For example, when discussing $\mathfrak{sl}_2$-modules in the next chapter we will use the notation e^2v for $e(ev)$; this has nothing to do with the square of the matrix e in Mat_2, which happens to be the zero matrix.

The obvious caveat associated with the module notation is that there can be more than one action of $\mathfrak{g}$ on the same vector space. It will occasionally be necessary to cope with this by introducing symbols for the representations.

Example 4.1.7. As well as the faithful action of $\mathfrak{l}_2$ on $\mathbb{C}^2$ considered in Example 4.1.4, there is a trivial action where both s and t act as 0. If we needed to discuss both, we would have to abandon the simplicity of writing $s_{\mathbb{C}^2}$ and sv, and instead write $\pi_i(s)$ and $\pi_i(s)v$, where $\pi_1 : \mathfrak{l}_2 \to \mathfrak{gl}(\mathbb{C}^2)$ was one representation and $\pi_2 : \mathfrak{l}_2 \to \mathfrak{gl}(\mathbb{C}^2)$ was the other. The question of what other actions $\mathfrak{l}_2$ has on $\mathbb{C}^2$ will be addressed, after a suitable reformulation, in Example 4.2.8 and Exercise 4.5.4, so we will say no more about it now; nevertheless, to give a small idea of how one can narrow down the search, we note here that if $t_{\mathbb{C}^2}$ is to equal the commutator $[s_{\mathbb{C}^2}, t_{\mathbb{C}^2}]$ then it must have trace zero. ■

This last observation is worth generalizing.

Proposition 4.1.8. *If V is a $\mathfrak{g}$-module and $x \in \mathcal{D}\mathfrak{g}$ then* $\mathrm{tr}(x_V) = 0$.

Proof. For any $y, z \in \mathfrak{g}$, $\mathrm{tr}([y,z]_V) = \mathrm{tr}([y_V, z_V]) = 0$. □

If $\mathfrak{g}$ is a linear Lie algebra, i.e. a subalgebra of some $\mathfrak{gl}_n$, then $\mathfrak{g}$ by definition has a representation on $\mathbb{C}^n$, namely the inclusion $\mathfrak{g} \to \mathfrak{gl}_n = \mathfrak{gl}(\mathbb{C}^n)$: every element of $\mathfrak{g}$ is already a matrix. This is usually called the *natural representation* of $\mathfrak{g}$, and $\mathbb{C}^n$ is called the *natural* $\mathfrak{g}$*-module*. In calculating with this module we will use the notation $v_1, \dots, v_n$ for the standard basis elements; the action of the matrices is determined by the rule

$$e_{ij}v_k = \delta_{jk}v_i. \tag{4.1.2}$$

Remark 4.1.9. The notion of a natural module of a linear Lie algebra $\mathfrak{g} \subseteq \mathfrak{gl}_n$ is not as intrinsic as it may seem, in that $\mathfrak{g}$ may be isomorphic to many different subalgebras of various $\mathfrak{gl}_m$. A trivial example is $\mathfrak{gl}_1 \cong \mathfrak{so}_2$, giving 'natural' representations of the one-dimensional Lie algebra both on $\mathbb{C}$ and on $\mathbb{C}^2$; another example is $\mathfrak{sl}_2 \cong \mathfrak{so}_3$, as seen in Exercise 2.3.5, giving 'natural' representations of this three-dimensional simple Lie algebra both on $\mathbb{C}^2$ and $\mathbb{C}^3$. Moreover, if τ is any automorphism of $\mathfrak{gl}_n$ then it restricts to give an isomorphism between $\mathfrak{g}$ and $\tau(\mathfrak{g})$ and there may be no canonical way to say which of these isomorphic copies 'is' the Lie algebra, for the purpose of defining its 'natural' representation.

Proposition 4.1.10. *If $\varphi : \mathfrak{g} \to \mathfrak{h}$ is a homomorphism of Lie algebras and V is an $\mathfrak{h}$-module then, using the action*

$$xv = \varphi(x)v \quad \text{for all } x \in \mathfrak{g},\ v \in V,$$

we can regard V as a $\mathfrak{g}$-module.

Proof. In terms of representations this simply says that when we compose the homomorphism $\varphi : \mathfrak{g} \to \mathfrak{h}$ with the homomorphism $\mathfrak{h} \to \mathfrak{gl}(V) : y \mapsto y_V$, we get another homomorphism $\mathfrak{g} \to \mathfrak{gl}(V)$. □

Example 4.1.11. One particular kind of homomorphism is the inclusion map of a subalgebra $\mathfrak{g}$ within a larger Lie algebra $\mathfrak{h}$. In this case the procedure in Proposition 4.1.10 is known as *restricting* the $\mathfrak{h}$-module V (or the action or representation of $\mathfrak{h}$ on V) to $\mathfrak{g}$. It is an advantage of the module notation that there is no difference between the action x_V whether x is viewed as an element of $\mathfrak{g}$ or as an element of $\mathfrak{h}$; the idea is that the element x has its own particular way of acting on the vector space V, namely by means of the linear transformation x_V. ■

Example 4.1.12. Another kind of homomorphism is the canonical projection from $\mathfrak{g}$ to a quotient Lie algebra $\mathfrak{h} = \mathfrak{g}/\mathfrak{g}'$ for some ideal $\mathfrak{g}'$ of $\mathfrak{g}$. In this case the procedure in Proposition 4.1.10 is known colloquially as *pulling back* the $\mathfrak{h}$-module V (or the action or representation of $\mathfrak{h}$ on V) to $\mathfrak{g}$. Notice that the ideal $\mathfrak{g}'$ is contained in the kernel of the resulting representation, which is therefore not faithful (assuming $\mathfrak{g}' \neq \{0\}$). In fact, every representation $\pi : \mathfrak{g} \to \mathfrak{gl}(V)$ is obtained by pulling back a faithful representation, namely the isomorphism $\mathfrak{g}/\ker(\pi) \xrightarrow{\sim} \pi(\mathfrak{g})$ provided by the fundamental homomorphism theorem. It may seem that π is 'really' a representation of the quotient $\mathfrak{g}/\ker(\pi)$ and not of $\mathfrak{g}$: why do we consider non-faithful representations at all? The benefits of allowing this extra flexibility in the definition of a representation will become clear as we proceed. ■

The most canonical representation of a Lie algebra is the *adjoint representation*, defined as follows.

Proposition 4.1.13. *Let $\mathfrak{g}$ be any Lie algebra. For any $x \in \mathfrak{g}$, define a linear transformation*

$$\mathrm{ad}_{\mathfrak{g}}(x) : \mathfrak{g} \to \mathfrak{g} : y \mapsto [x, y].$$

Then $\mathrm{ad}_{\mathfrak{g}} : \mathfrak{g} \to \mathfrak{gl}(\mathfrak{g})$ is a representation of $\mathfrak{g}$ on $\mathfrak{g}$ itself.

Proof. It is clear that $\mathrm{ad}_{\mathfrak{g}}$ is a linear map. We need to show that $\mathrm{ad}_{\mathfrak{g}}([x, y]) = [\mathrm{ad}_{\mathfrak{g}}(x), \mathrm{ad}_{\mathfrak{g}}(y)]$ for all $x, y \in \mathfrak{g}$, i.e. that

$$[[x, y], z] = [x, [y, z]] - [y, [x, z]] \quad \text{for all } x, y, z \in \mathfrak{g}. \tag{4.1.3}$$

This is, however, just a form of the Jacobi identity. □

Use of the notation $x_{\mathfrak{g}}$ for the adjoint representation could lead to confusion so we will keep to $\mathrm{ad}_{\mathfrak{g}}(x)$, abbreviated to $\mathrm{ad}(x)$ if the Lie algebra is determined by the context. Note that the kernel of the adjoint representation is, by definition, the centre $Z(\mathfrak{g})$; in particular, $\mathfrak{g}$ is a faithful $\mathfrak{g}$-module if and only if $Z(\mathfrak{g}) = \{0\}$.

Example 4.1.14. In the adjoint representation of $\mathfrak{sl}_2$, the element e is represented by the linear transformation $\mathrm{ad}(e)$ of $\mathfrak{sl}_2$, which sends any element x to $[e, x]$. On the standard basis, this linear transformation acts as follows:

$$\mathrm{ad}(e)e = [e, e] = 0, \quad \mathrm{ad}(e)h = [e, h] = -2e, \quad \mathrm{ad}(e)f = [e, f] = h.$$

Making similar calculations for $\mathrm{ad}(h)$ and $\mathrm{ad}(f)$, we deduce that the representing matrices are:

$$\mathrm{ad}(e) : \begin{pmatrix} 0 & -2 & 0 \\ 0 & 0 & 1 \\ 0 & 0 & 0 \end{pmatrix}, \quad \mathrm{ad}(h) : \begin{pmatrix} 2 & 0 & 0 \\ 0 & 0 & 0 \\ 0 & 0 & -2 \end{pmatrix}, \quad \mathrm{ad}(f) : \begin{pmatrix} 0 & 0 & 0 \\ -1 & 0 & 0 \\ 0 & 2 & 0 \end{pmatrix}.$$

Since $Z(\mathfrak{sl}_2) = \{0\}$ (or alternatively because these matrices are clearly linearly independent), this is a faithful representation of $\mathfrak{sl}_2$. ■

To conclude this section we treat the easy case of one-dimensional modules. If V is a one-dimensional vector space then $\mathfrak{gl}(V)$ is a one-dimensional Lie algebra and can be identified with $\mathbb{C}$ in a canonical way (every linear transformation of V is just multiplication by some $a \in \mathbb{C}$; its matrix is the 1×1 matrix with a as its sole entry). So discussing one-dimensional $\mathfrak{g}$-modules amounts to discussing Lie algebra homomorphisms $\mathfrak{g} \to \mathbb{C}$. There is a special name for these.

Definition 4.1.15. If $\mathfrak{g}$ is a Lie algebra, a *character of* $\mathfrak{g}$ is a Lie algebra homomorphism $\chi : \mathfrak{g} \to \mathbb{C}$ or, in other words, a linear function on $\mathfrak{g}$ that satisfies $\chi([x, y]) = 0$ for all $x, y \in \mathfrak{g}$.

Proposition 4.1.16. *The characters of $\mathfrak{g}$ are precisely the linear functions on $\mathfrak{g}$ obtained by pulling back linear functions on the maximal abelian quotient $\mathfrak{g}/\mathcal{D}\mathfrak{g}$.*

Proof. Since $\mathcal{D}\mathfrak{g}$ is the span of all brackets $[x, y]$ for $x, y \in \mathfrak{g}$, the condition for a linear function χ on $\mathfrak{g}$ to be a character is equivalent to saying that χ vanishes on $\mathcal{D}\mathfrak{g}$. The result follows. □

Example 4.1.17. If $\mathfrak{g}$ is itself abelian then the characters of $\mathfrak{g}$ are just the linear functions on $\mathfrak{g}$, i.e. the elements of the dual space $\mathfrak{g}^*$. ■

Example 4.1.18. Since $\mathcal{D}\mathfrak{sl}_n = \mathfrak{sl}_n$, the only character of $\mathfrak{sl}_n$ is the zero function. ■

Example 4.1.19. The trace function $\mathrm{tr} : \mathfrak{gl}_n \to \mathbb{C}$ is clearly a character of $\mathfrak{gl}_n$. Since $\mathcal{D}\mathfrak{gl}_n = \mathfrak{sl}_n$ has codimension 1 in $\mathfrak{gl}_n$, the characters of $\mathfrak{gl}_n$ form a one-dimensional vector space, so they are precisely the scalar multiples of the trace. ■

Definition 4.1.20. For any Lie algebra $\mathfrak{g}$ and character $\chi : \mathfrak{g} \to \mathbb{C}$, we write $\mathbb{C}_\chi$ for the $\mathfrak{g}$-module that as a vector space is $\mathbb{C}$, with representation χ; that is, any $x \in \mathfrak{g}$ acts as multiplication by $\chi(x)$. If $\chi = 0$ then we omit the subscript and write simply $\mathbb{C}$ for this *trivial module*, where all elements of $\mathfrak{g}$ act as zero.

4.2 Isomorphism of modules

For each Lie algebra $\mathfrak{g}$, the $\mathfrak{g}$-modules form a linear category in the sense that there is a concept of homomorphisms between them that behaves much like the concept of linear maps between vector spaces.

Definition 4.2.1. Let V and W be $\mathfrak{g}$-modules. A *$\mathfrak{g}$-module homomorphism* or *$\mathfrak{g}$-linear map* from V to W is a linear map $\varphi : V \to W$ such that

$$\varphi(xv) = x\varphi(v) \quad \text{for all } x \in \mathfrak{g},\ v \in V.$$

The set of such $\mathfrak{g}$-module homomorphisms is written $\mathrm{Hom}_{\mathfrak{g}}(V, W)$; it is clearly a vector space. In the special case where $W = V$, a $\mathfrak{g}$-module homomorphism from V to V is called a *$\mathfrak{g}$-module endomorphism* of V. If $\varphi \in \mathrm{Hom}_{\mathfrak{g}}(V, W)$ is invertible, it is called an *isomorphism*; its inverse is obviously another isomorphism. If such an isomorphism exists, we will say that the $\mathfrak{g}$-modules V and W are *isomorphic*, written $V \cong W$, and that the representations of $\mathfrak{g}$ on V and on W are *equivalent*.

Proposition 4.2.2. *Two $\mathfrak{g}$-modules V and W are isomorphic if and only if there are bases of V and W with respect to which the matrices representing any given element of $\mathfrak{g}$ are the same.*

Proof. This follows immediately from the definitions. □

Of course, $\mathfrak{g}$-modules of different dimensions cannot be isomorphic.

Example 4.2.3. Any n-dimensional $\mathfrak{g}$-module V is isomorphic to one whose underlying vector space is $\mathbb{C}^n$. Indeed, when we find the representing matrices relative to a chosen basis of V, what we are doing is converting the representation $\mathfrak{g} \to \mathfrak{gl}(V)$ into an equivalent representation $\mathfrak{g} \to \mathfrak{gl}_n = \mathfrak{gl}(\mathbb{C}^n)$. ■

Example 4.2.4. As a special case of the previous example, every one-dimensional $\mathfrak{g}$-module V is isomorphic to $\mathbb{C}_\chi$ for some character $\chi : \mathfrak{g} \to \mathbb{C}$. In fact χ is uniquely determined by V, because $\chi(x)$ is the scalar by which x acts on V. ■

Example 4.2.5. Consider the adjoint representation of $\mathfrak{l}_2$: $\mathrm{ad}(s)$ annihilates s and fixes t, and $\mathrm{ad}(t)$ sends s to $-t$ and annihilates t. Thus, with respect to the standard basis, the matrix representing s is $\left(\begin{smallmatrix} 0 & 0 \\ 0 & 1 \end{smallmatrix}\right)$ and that representing t is $\left(\begin{smallmatrix} 0 & 0 \\ -1 & 0 \end{smallmatrix}\right)$. But if we change the basis of the vector space to $t, -s$ then the matrix representing s becomes $\left(\begin{smallmatrix} 1 & 0 \\ 0 & 0 \end{smallmatrix}\right)$ and that representing t becomes $\left(\begin{smallmatrix} 0 & 1 \\ 0 & 0 \end{smallmatrix}\right)$, as in Example 4.1.4. So the $\mathfrak{l}_2$-module $\mathfrak{l}_2$ is isomorphic to the $\mathfrak{l}_2$-module $\mathbb{C}^2$ defined in Example 4.1.4. ■

Two different $\mathfrak{g}$-actions on the same vector space V can give isomorphic modules. According to Definition 4.2.1, two representations $\pi_1, \pi_2 : \mathfrak{g} \to \mathfrak{gl}(V)$ are equivalent if and only if there is some invertible linear transformation φ of V such that

$$\pi_2(x) = \varphi \pi_1(x) \varphi^{-1} \quad \text{for all } x \in \mathfrak{g}. \tag{4.2.1}$$

In terms of the representing matrices relative to a chosen basis of V, the above condition amounts to saying that there is a single invertible matrix that conjugates one whole collection of representing matrices to the other.

Example 4.2.6. The two isomorphic $\mathfrak{l}_2$-modules discussed in Example 4.2.5 could have been defined using the same underlying vector space, say $\mathbb{C}^2$; it was just notationally more convenient to use $\mathbb{C}^2$ as the vector space for the first and $\mathfrak{l}_2$ as the vector space for the second. Whatever the vector spaces, all that matters for the isomorphism or equivalence is that the two collections of representing matrices, namely

$$\left(s : \begin{pmatrix} 1 & 0 \\ 0 & 0 \end{pmatrix}, t : \begin{pmatrix} 0 & 1 \\ 0 & 0 \end{pmatrix}\right) \quad \text{and} \quad \left(s : \begin{pmatrix} 0 & 0 \\ 0 & 1 \end{pmatrix}, t : \begin{pmatrix} 0 & 0 \\ -1 & 0 \end{pmatrix}\right),$$

are related by a 'simultaneous conjugation', namely that

$$g \begin{pmatrix} 1 & 0 \\ 0 & 0 \end{pmatrix} g^{-1} = \begin{pmatrix} 0 & 0 \\ 0 & 1 \end{pmatrix} \quad \text{and} \quad g \begin{pmatrix} 0 & 1 \\ 0 & 0 \end{pmatrix} g^{-1} = \begin{pmatrix} 0 & 0 \\ -1 & 0 \end{pmatrix}$$

for some $g \in GL_2$: one choice is $g = \left(\begin{smallmatrix} 0 & 1 \\ -1 & 0 \end{smallmatrix}\right)$. ■

To study the *representation theory* of a Lie algebra $\mathfrak{g}$ is to study the category of $\mathfrak{g}$-modules, with the aim of tackling problems such as the following, or special cases of it:

Problem 4.2.7. For a given $\mathfrak{g}$, classify the $\mathfrak{g}$-modules up to isomorphism. In other words, classify representations of $\mathfrak{g}$ up to equivalence.

For instance, when $\mathfrak{g} = \mathfrak{gl}_n$ this is a more precise formulation of the vague Problem 1.3.3.

Example 4.2.8. To illustrate Problem 4.2.7, suppose that we want to classify up to isomorphism all two-dimensional $\mathfrak{l}_2$-modules V in which t_V is nonzero and nilpotent (i.e. its only eigenvalue is zero). At this stage it is more convenient to use the alternative viewpoint, in which the problem is to classify up to equivalence all representations π of $\mathfrak{l}_2$ on $\mathbb{C}^2$ in which $\pi(t)$ is nonzero and nilpotent. By the Jordan canonical form theorem, $\pi(t)$ must be similar to the matrix $\left(\begin{smallmatrix} 0 & 1 \\ 0 & 0 \end{smallmatrix}\right)$, so (passing to an equivalent representation if necessary) we can assume that $\pi(t)$ equals this matrix. If we let $\left(\begin{smallmatrix} a & b \\ c & d \end{smallmatrix}\right)$ denote the matrix of $\pi(s)$, we must have the bracket relation

$$\begin{pmatrix} 0 & 1 \\ 0 & 0 \end{pmatrix} = \left[\begin{pmatrix} a & b \\ c & d \end{pmatrix}, \begin{pmatrix} 0 & 1 \\ 0 & 0 \end{pmatrix}\right] = \begin{pmatrix} -c & a-d \\ 0 & c \end{pmatrix},$$

so that $c = 0$ and $d = a - 1$. Conversely, setting

$$\pi_{a,b}(s) = \begin{pmatrix} a & b \\ 0 & a-1 \end{pmatrix} \quad \text{and} \quad \pi_{a,b}(t) = \begin{pmatrix} 0 & 1 \\ 0 & 0 \end{pmatrix}$$

defines a representation $\pi_{a,b}$ of $\mathfrak{l}_2$ on $\mathbb{C}^2$ for any $a, b \in \mathbb{C}$. The remaining question is, when is $\pi_{a,b}$ equivalent to $\pi_{c,d}$? To answer this, we have to decide when the equations

$$g \begin{pmatrix} a & b \\ 0 & a-1 \end{pmatrix} g^{-1} = \begin{pmatrix} c & d \\ 0 & c-1 \end{pmatrix} \quad \text{and} \quad g \begin{pmatrix} 0 & 1 \\ 0 & 0 \end{pmatrix} g^{-1} = \begin{pmatrix} 0 & 1 \\ 0 & 0 \end{pmatrix}$$

have a simultaneous solution for $g \in GL_2$. For the first equation to have a solution, the eigenvalues $a, a-1$ of one matrix have to equal the eigenvalues $c, c-1$ of the other, so a has to equal c; we will assume this henceforth. By direct calculation the solutions of the second equation (i.e. the invertible matrices commuting with $\left(\begin{smallmatrix} 0 & 1 \\ 0 & 0 \end{smallmatrix}\right)$) are those matrices g of the form $\left(\begin{smallmatrix} \alpha & \beta \\ 0 & \alpha \end{smallmatrix}\right)$, where $\alpha \in \mathbb{C}^\times$, $\beta \in \mathbb{C}$. Substituting this into the first equation, we obtain

$$\alpha b + \beta(a-1) = \beta a + \alpha d,$$

which is automatically true if β is chosen to be $\alpha(b-d)$ So $\pi_{a,b}$ is equivalent to $\pi_{a,d}$ for any d but is not equivalent to $\pi_{c,d}$, where $c \neq a$. Thus we have a one-parameter family of equivalence classes, where the parameter a has the interpretation of being the larger of the eigenvalues of the matrix representing s, the other eigenvalue being necessarily $a-1$. The general classification of two-dimensional $\mathfrak{l}_2$-modules is left to Exercise 4.5.4. ∎

4.3 Submodules and irreducible modules

In tackling Problem 4.2.7 the direct matrix-based approach of Example 4.2.8 is found to be unsuitable for larger dimensions. As usual, a better approach is to look at the internal structure of a $\mathfrak{g}$-module.

Definition 4.3.1. Let V be a $\mathfrak{g}$-module. A $\mathfrak{g}$-*submodule*, or just *submodule* (also known as an *invariant subspace*), of V is a subspace $W \subseteq V$ such that $xW \subseteq W$ for all $x \in \mathfrak{g}$. It is thus a $\mathfrak{g}$-module in its own right, with $x_W = x_V|_W$. The corresponding *quotient module* is V/W, which is a $\mathfrak{g}$-module with $x_{V/W} = x_V|_{V/W}$. A nonzero $\mathfrak{g}$-module V is *irreducible* or *simple* if it has no submodules except the *trivial submodules* $\{0\}$ and V and is *reducible* if it does have a nontrivial submodule. The terms 'irreducible' and 'reducible' are also applied to the representation of $\mathfrak{g}$ on V.

Note that the submodules and irreducibility of a $\mathfrak{g}$-module V depend only on the Lie algebra $\mathfrak{g}_V$ through which $\mathfrak{g}$ actually acts; they cannot detect anything about the kernel of the representation. In particular, if V is a $\mathfrak{g}$-module obtained by pulling back a module over some quotient $\mathfrak{h}$ of $\mathfrak{g}$ then its submodules as a $\mathfrak{g}$-module are the same as its submodules as an $\mathfrak{h}$-module and it is irreducible as a $\mathfrak{g}$-module if and only if it is irreducible as an $\mathfrak{h}$-module.

Example 4.3.2. A one-dimensional module is automatically irreducible. ∎

Example 4.3.3. In the $\mathfrak{l}_2$-module $\mathbb{C}^2$ (with the action of Example 4.1.4), $\mathbb{C}v_1$ is an $\mathfrak{l}_2$-submodule. So $\mathbb{C}^2$ is a reducible $\mathfrak{l}_2$-module. ∎

Submodules are related to homomorphisms in the same way as in the theory of modules over rings.

Proposition 4.3.4. *Let $\varphi : V \to W$ be a $\mathfrak{g}$-module homomorphism. Then*

(1) $\ker(\varphi)$ *is a submodule of* V;
(2) $\operatorname{im}(\varphi)$ *is a submodule of* W;
(3) *the natural map* $V/\ker(\varphi) \to \operatorname{im}(\varphi) : v + \ker(\varphi) \mapsto \varphi(v)$ *is a* $\mathfrak{g}$-*module isomorphism.*

Proof. This all follows easily from the relevant definitions. □

In a general $\mathfrak{g}$-module V, a one-dimensional subspace $\mathbb{C}v$ is a submodule if and only if $xv \in \mathbb{C}v$ for all $x \in \mathfrak{g}$; in other words, v is an eigenvector for every x_V. If this is the case then, by Example 4.2.4, there is a unique character $\chi : \mathfrak{g} \to \mathbb{C}$ such that $xv = \chi(x)v$ for all $x \in \mathfrak{g}$; in other words, the eigenvalue with respect to which v is an eigenvector for x_V is exactly $\chi(x)$. So the character χ prescribes the different eigenvalues by which different elements of the Lie algebra act on the common eigenvector v.

Definition 4.3.5. If V is a $\mathfrak{g}$-module and $\chi : \mathfrak{g} \to \mathbb{C}$ is a character, the *χ-eigenspace of $\mathfrak{g}$ in V*, written $V_\chi^\mathfrak{g}$, is defined to be

$$\{v \in V \,|\, xv = \chi(x)v \text{ for all } x \in \mathfrak{g}\}.$$

The nonzero elements of $V_\chi^\mathfrak{g}$ are called *χ-eigenvectors of $\mathfrak{g}$ in V*; these are precisely the elements of V whose span is a one-dimensional submodule isomorphic to $\mathbb{C}_\chi$. If $\chi = 0$, we omit the subscript and write $V^\mathfrak{g}$ for $\{v \in V \,|\, xv = 0 \text{ for all } x \in \mathfrak{g}\}$. Instead of 0-eigenvectors, we call the elements of $V^\mathfrak{g}$ the *invariant vectors of $\mathfrak{g}$ in V*.

Remark 4.3.6. The reason for describing elements of $V^\mathfrak{g}$ as invariant when in fact they are annihilated by the action of $\mathfrak{g}$, is that they are invariant under elements of the corresponding Lie group: the condition $xv = 0$ is a 'differentiated' version of the condition $gv = v$.

Clearly each eigenspace $V_\chi^\mathfrak{g}$ is a submodule of V. Note that if $\mathfrak{g} = \mathbb{C}x$ is one-dimensional then $V_\chi^\mathfrak{g}$ is the same as the $\chi(x)$-eigenspace of x_V, in the terminology of ordinary linear algebra. In general, if $\mathfrak{g}$ has a basis $x_1, \ldots, x_m$ then $V_\chi^\mathfrak{g}$ is the intersection over all i of the $\chi(x_i)$-eigenspace of $(x_i)_V$. The next two propositions generalize standard results about the eigenvectors of a single linear transformation to the case of a whole Lie algebra.

Proposition 4.3.7. *If V is a $\mathfrak{g}$-module, the sum of all the eigenspaces is the direct sum $\bigoplus_\chi V_\chi^\mathfrak{g}$. In other words, any collection of eigenvectors for distinct characters of $\mathfrak{g}$ must be linearly independent.*

Proof. Suppose for a contradiction that the sum of the eigenspaces is not direct. Then there are distinct characters $\chi_1, \ldots, \chi_s$ of $\mathfrak{g}$ and corresponding eigenvectors $v_1, \ldots, v_s$ in V such that $v_1 + \cdots + v_s = 0$. We can assume that s is minimal for such an equation to hold; clearly $s \geq 2$. Let $x \in \mathfrak{g}$ be such that $\chi_1(x) \neq \chi_2(x)$. Then

$$\begin{aligned} 0 &= x(v_1 + v_2 + \cdots + v_s) - \chi_1(x)(v_1 + v_2 + \cdots + v_s) \\ &= (\chi_2(x) - \chi_1(x))\, v_2 + \cdots + (\chi_s(x) - \chi_1(x))\, v_s. \end{aligned}$$

Now each term $(\chi_i(x) - \chi_1(x))v_i$ is either zero or a χ_i-eigenvector of $\mathfrak{g}$, and $(\chi_2(x) - \chi_1(x))v_2$ is definitely nonzero. So we have a contradiction to the minimality of s. ☐

Proposition 4.3.8. *If $\mathfrak{g}$ is an abelian Lie algebra and V is a nonzero $\mathfrak{g}$-module then $V_\chi^\mathfrak{g} \neq 0$ for some character χ of $\mathfrak{g}$. In other words, V contains a one-dimensional $\mathfrak{g}$-submodule. So the only irreducible $\mathfrak{g}$-modules are those that are one-dimensional.*

Proof. The proposition essentially says that a collection of commuting linear transformations must have a common eigenvector. Let $x_1, \dots, x_m$ be a basis of $\mathfrak{g}$. By basic linear algebra, $(x_1)_V$ must have some nonzero eigenspace, say W_1. Since $[(x_1)_V, (x_i)_V] = 0$ for all i, it follows that W_1 is preserved by all $(x_i)_V$, using the same argument as in the proof of Proposition 3.2.10. So W_1 is a nonzero $\mathfrak{g}$-submodule of V consisting entirely of eigenvectors for $(x_1)_V$ (and the zero vector). Then $(x_2)_{W_1}$ must have some nonzero eigenspace, say W_2, that is also a $\mathfrak{g}$-submodule, by similar reasoning, and consists entirely of common eigenvectors for $(x_1)_V$ and $(x_2)_V$ (and the zero vector). Continuing in this way, we eventually find a nonzero $\mathfrak{g}$-submodule W_m of V that consists entirely of eigenvectors of $\mathfrak{g}$ (and the zero vector). Any one-dimensional subspace of W_m is a one-dimensional $\mathfrak{g}$-submodule of V. □

Remark 4.3.9. If you have experience of group representations, you will see an analogy between Proposition 4.3.8 and the case of abelian groups.

Finding submodules of dimension greater than 1 is usually harder than the one-dimensional case, where it just amounts to identifying common eigenvectors. It can be useful to bear in mind the following way of visualizing submodules in terms of representing matrices. If V is an n-dimensional $\mathfrak{g}$-module then, for any k-dimensional subspace $W \subseteq V$, there is a basis $v_1, \dots, v_n$ of V such that $W = \mathbb{C}\{v_1, \dots, v_k\}$. With respect to such a basis, W is a $\mathfrak{g}$-submodule if and only if the matrices representing the elements of $\mathfrak{g}$ lie in the 'block upper-triangular' subalgebra $\left(\begin{smallmatrix} * & * \\ 0 & * \end{smallmatrix}\right)$, with an $(n-k) \times k$ block of zeroes.

Example 4.3.10. The natural $\mathfrak{gl}_n$-module $\mathbb{C}^n$ is irreducible, because (as already mentioned in the proof of Proposition 3.2.10) there is no nontrivial subspace of $\mathbb{C}^n$ which is preserved by every matrix. When you restrict this module to a subalgebra $\mathfrak{g}$ of $\mathfrak{gl}_n$, it may become reducible or it may not. For instance, when $n \geq 2$, $\mathbb{C}^n$ is a reducible $\mathfrak{b}_n$-module because the span of $v_1, \dots, v_k$ is a $\mathfrak{b}_n$-submodule for any k, as is clear from the upper triangularity of the matrices. By contrast, $\mathbb{C}^n$ is an irreducible $\mathfrak{sl}_n$-module because any subspace that is preserved by all $x \in \mathfrak{sl}_n$ must be preserved by all $x + a1_n$ for $x \in \mathfrak{sl}_n$ and $a \in \mathbb{C}$, i.e. every element of $\mathfrak{gl}_n$. ■

Remark 4.3.11. In general, if $\mathfrak{g}$ is a subalgebra of $\mathfrak{gl}_n$ that generates the associative algebra Mat_n (i.e. when you allow actual matrix multiplication, not just commutators, together with linear combinations you can generate any matrix from the elements of $\mathfrak{g}$) then any subspace preserved by all $x \in \mathfrak{g}$ must be preserved by every matrix, so in this case $\mathbb{C}^n$ is an irreducible $\mathfrak{g}$-module. By a standard result in ring theory known as the density theorem, the converse to this statement also holds: if $\mathfrak{g}$

is a subalgebra of $\mathfrak{gl}_n$ such that $\mathbb{C}^n$ is an irreducible $\mathfrak{g}$-module then $\mathfrak{g}$ must generate the associative algebra Mat_n.

Example 4.3.12. A $\mathfrak{g}$-submodule of $\mathfrak{g}$ (with the adjoint representation) is precisely an ideal of $\mathfrak{g}$. The adjoint representation of $\mathfrak{g}$ is irreducible exactly when $\mathfrak{g}$ is either simple or one-dimensional. ■

Clearly the sum of a collection of $\mathfrak{g}$-submodules of a $\mathfrak{g}$-module V is also a $\mathfrak{g}$-submodule, as is their intersection. If S is a subset of V, we can define the $\mathfrak{g}$-submodule *generated by* S to be the intersection of all submodules containing S.

Proposition 4.3.13. *The $\mathfrak{g}$-submodule of V generated by S equals the span of all elements $x_1x_2 \cdots x_ts$ for $t \geq 0$, $x_1, \dots, x_t \in \mathfrak{g}$, $s \in S$.*

Proof. This span is clearly contained in any $\mathfrak{g}$-submodule that contains S. It is also easy to see that the span is a $\mathfrak{g}$-submodule, whence the result. □

Proposition 4.3.14. *A nonzero $\mathfrak{g}$-module V is irreducible if and only if, for all nonzero $v \in V$, the submodule generated by v is V.*

Proof. If V is irreducible then the submodule generated by any nonzero v must be the whole of V, since that is the only nonzero submodule. Conversely, suppose that every nonzero element generates the whole module: then there cannot be a nontrivial submodule W of V, because any nonzero element of W would have to generate a submodule of W. □

Note that Proposition 4.3.14 would become false if one changed 'for all nonzero $v \in V$' to 'for some nonzero $v \in V$'.

Example 4.3.15. We have seen that $\mathbb{C}^2$ is reducible as a $\mathfrak{b}_2$-module, but it is also generated by a single nonzero element, namely the second standard basis element v_2, because $v_1 = e_{12}v_2$. ■

4.4 Complete reducibility

If a $\mathfrak{g}$-module V has a nontrivial submodule W then V can be regarded as being built from the $\mathfrak{g}$-modules W and V/W; these in turn are built from smaller modules, and so there is a loose sense in which any module is built from irreducible modules. The difficulty comes in describing exactly how the pieces fit together.

Although it is far from being the whole story, there is an easy way to combine two modules to form a larger module, namely the direct sum.

Proposition 4.4.1. *If W and W' are $\mathfrak{g}$-modules, the direct sum $W \oplus W'$ is a $\mathfrak{g}$-module under the $\mathfrak{g}$-action defined by $x(v, v') = (xv, xv')$. If V is a $\mathfrak{g}$-module* *that, as a vector space, is the direct sum of two subspaces W and W', and these are $\mathfrak{g}$-submodules of V, then the vector space isomorphism $W \oplus W' \to V : (v, v') \mapsto v + v'$ is a $\mathfrak{g}$-module isomorphism.*

Proof. This follows immediately from the relevant definitions. □

The same principles apply to the direct sum of more than two modules.

Example 4.4.2. The $\mathfrak{d}_n$-module $\mathbb{C}^n$ is the direct sum of one-dimensional submodules: $\mathbb{C}^n = \mathbb{C}v_1 \oplus \cdots \oplus \mathbb{C}v_n$. ■

Example 4.4.3. Returning to the $\mathfrak{b}_2$-module $\mathbb{C}^2$ we see that, despite being reducible, it cannot be written as a direct sum of smaller $\mathfrak{b}_2$-modules: there is only a single one-dimensional $\mathfrak{b}_2$-submodule because there is (up to a multiplicative scalar) only one common eigenvector of the matrices of $\mathfrak{b}_2$, namely v_1. This module therefore fails to be completely reducible in the sense of the next definition. ■

Definition 4.4.4. We say that a $\mathfrak{g}$-module V is *completely reducible* or *semisimple* if, for every submodule W, V contains a *complementary* submodule W', i.e. one satisfying $V = W \oplus W'$.

Theorem 4.4.5. *The $\mathfrak{g}$-module V is completely reducible if and only if V can be written as a direct sum of irreducible submodules.*

Proof. First we assume that V is completely reducible; we are aiming to show that V is a direct sum of irreducible submodules. If $V = \{0\}$ then this is vacuously true (the zero vector space is an empty direct sum). If V is nonzero and irreducible then it is true because V equals itself (it is a direct sum of one module). Otherwise, V has a nontrivial submodule W and by assumption there is a submodule W' such that $V = W \oplus W'$. Now, W is itself completely reducible because if X is a submodule of W (and hence also of V) then there is a submodule X' of V such that $V = X \oplus X'$, and it is easy to see that the submodule $X' \cap W$ of W satisfies $W = X \oplus (X' \cap W)$. Similarly, W' is completely reducible. Since W and W' both have dimensions smaller than V, we can assume by induction that they can be written as direct sums of irreducible submodules, and hence V can also be written thus.

Conversely, suppose that V is a direct sum of irreducible submodules; we want to show that V is completely reducible. If $V = \{0\}$ or V is irreducible, this is trivially true. Otherwise, we have $V = V_1 \oplus V_2$ where V_1 is an irreducible submodule and V_2 is a nonzero direct sum of irreducible submodules. Since V_2 has a smaller dimension than V, we can assume by induction that V_2 is completely reducible. Now let W be any submodule of V. The intersection $W \cap V_1$ is a submodule of V_1, which must be either $\{0\}$ or V_1 because V_1 is irreducible. Define $\widetilde{W}$ to equal $W \oplus V_1$ if $W \cap V_1 = \{0\}$

and to equal W otherwise; thus we have $\widetilde{W} \supseteq V_1$ in either case. Now $\widetilde{W} \cap V_2$ is a submodule of V_2, and since V_2 is completely reducible there must be another submodule W' such that $V_2 = (\widetilde{W} \cap V_2) \oplus W'$. It is then straightforward to deduce that $V = \widetilde{W} \oplus W'$. So either $V_1 \oplus W'$ (in the case $W \cap V_1 = \{0\}$) or W' (in the case $W \cap V_1 = V_1$) is the desired submodule complementary to W in V. □

Example 4.4.6. Consider the smallest nonzero Lie algebra, $\mathfrak{g} = \mathbb{C}x$. From what we have seen, the isomorphism classes of n-dimensional $\mathbb{C}x$-modules are in bijection with the similarity classes of $n \times n$ matrices (to the $\mathbb{C}x$-module V corresponds the matrix representing x on V), so they are classified by the Jordan canonical form theorem. Since a $\mathbb{C}x$-module is irreducible if and only if it is one-dimensional, the completely reducible $\mathbb{C}x$-modules correspond to diagonalizable matrices, those for which $\mathbb{C}^n$ has a basis of eigenvectors. A non-diagonalizable matrix gives a module that is not completely reducible: for example, in the $\mathbb{C}x$-module $\mathbb{C}^2$ defined by the matrix $\left(\begin{smallmatrix} 1 & 1 \\ 0 & 1 \end{smallmatrix}\right)$ there is no submodule complementary to the submodule $\mathbb{C}v_1$. ■

Example 4.4.7. More generally, if $\mathfrak{g}$ is any Lie algebra such that $\mathcal{D}\mathfrak{g}$ is not the whole of $\mathfrak{g}$ then there must exist $\mathfrak{g}$-modules that are not completely reducible. This follows immediately from the previous example: if $\mathfrak{h}$ is any codimension-1 subspace of $\mathfrak{g}$ containing $\mathcal{D}\mathfrak{g}$ then the quotient Lie algebra $\mathfrak{g}/\mathfrak{h}$ is one-dimensional, and we can pull back a non-completely-reducible module for $\mathfrak{g}/\mathfrak{h}$ to get a non-completely-reducible module for $\mathfrak{g}$. This gives us, for instance, non-completely-reducible $\mathfrak{gl}_n$-modules for any n. ■

Remark 4.4.8. Readers familiar with the representation theory of finite groups, where complete reducibility is guaranteed by Maschke's theorem, may be surprised that complete reducibility fails for modules over most Lie algebras. But it must be remembered that Lie algebras correspond to infinite groups. For instance, to involve the matrix $\left(\begin{smallmatrix} 1 & 1 \\ 0 & 1 \end{smallmatrix}\right) \in GL_2$ in a group representation, one has to consider an infinite group because this matrix has infinite order.

In the presence of non-complete reducibility, there can often be great subtlety in how irreducible modules fit together to make arbitrary modules. Consequently, classifying arbitrary modules can be difficult even in the case of abelian Lie algebras, where classifying irreducible modules is easy. In the case of a one-dimensional Lie algebra, we noted in Example 4.4.6 that we can classify the modules up to isomorphism, using Jordan canonical form. For a k-dimensional abelian Lie algebra, however, the general classification problem is equivalent to classifying k-tuples of commuting matrices up to simultaneous conjugation, which is notoriously intractable once k becomes large. The bad news is that this problem is not restricted to abelian Lie algebras: by the same sort of pulling back as in Example 4.4.7, the representation theory of an arbitrary Lie algebra $\mathfrak{g}$ has to be at least as complicated as that of its maximal abelian quotient $\mathfrak{g}/\mathcal{D}\mathfrak{g}$.

Now for the good news. Recall that if $\mathfrak{g}$ is a simple Lie algebra then $\mathfrak{g}/\mathcal{D}\mathfrak{g}$ is zero. It turns out that, for simple $\mathfrak{g}$, all modules are completely reducible, a result known as Weyl's theorem. Consequently, the representation theory more or less reduces to the study of irreducible modules, and the irreducible modules can be classified in a natural and relatively explicit way. For $\mathfrak{sl}_n$, these results are Theorems 7.4.9 and 7.5.3; for other simple Lie algebras, see Section 8.2.

Where does this leave our goal of classifying $\mathfrak{gl}_n$-modules? Since the difference between $\mathfrak{gl}_n$ and $\mathcal{D}\mathfrak{gl}_n = \mathfrak{sl}_n$ is only one dimension, the non-complete reducibility in the theory of $\mathfrak{gl}_n$-modules is manageable. In any case, however, our motivating statement, Problem 1.1.1, gave rise to a natural extra restriction, which we incorporated into Problem 1.4.2 and can revive here:

Definition 4.4.9. We say that a $\mathfrak{gl}_n$-module V is *integral* if $(e_{ii})_V$ is diagonalizable with integer eigenvalues, for all i.

We will see in Corollary 7.5.4 that integrality implies complete reducibility and will thereby solve the following precise version of Problem 1.4.2:

Problem 4.4.10. Classify integral $\mathfrak{gl}_n$-modules up to isomorphism.

4.5 Exercises

Exercise 4.5.1. Suppose that we have an $\mathfrak{l}_2$-action on $\mathbb{C}^3$ for which the representing matrices are

$$s_{\mathbb{C}^3} : \begin{pmatrix} 1 & 0 & 0 \\ a & b & 0 \\ c & d & e \end{pmatrix}, \qquad t_{\mathbb{C}^3} : \begin{pmatrix} 0 & 1 & 0 \\ 0 & 0 & 1 \\ 0 & 0 & 0 \end{pmatrix}.$$

(i) Determine a, b, c, d, e.
(ii) Let $v \in \mathbb{C}^3$. What are the possibilities for the $\mathfrak{l}_2$-submodule of $\mathbb{C}^3$ generated by v?

Exercise 4.5.2. Restrict the Lie algebra homomorphism $\psi : \mathfrak{gl}_2 \to \mathfrak{gl}_3$ of Example 1.3.1 to $\mathfrak{sl}_2$, and regard it as a representation of $\mathfrak{sl}_2$ on $\mathbb{C}^3$. Find an explicit $\mathfrak{sl}_2$-module isomorphism $\mathfrak{sl}_2 \xrightarrow{\sim} \mathbb{C}^3$.

Exercise 4.5.3. Regard $\mathfrak{gl}_2$ as a $\mathfrak{b}_2$-module by restricting the adjoint representation of $\mathfrak{gl}_2$ to $\mathfrak{b}_2$.

(i) Write down the representing matrices for $e_{11}, e_{12}, e_{22} \in \mathfrak{b}_2$ with respect to the usual basis of $\mathfrak{gl}_2$.

(ii) Show that $\mathfrak{gl}_2$ is a reducible $\mathfrak{b}_2$-module.
(iii) Is $\mathfrak{gl}_2$ a completely reducible $\mathfrak{b}_2$-module?

Exercise 4.5.4. (i) Classify two-dimensional $\mathfrak{l}_2$-modules up to isomorphism. (Part of this was done in Example 4.2.8).
(ii) Let $\mathfrak{g}$ denote a two-dimensional abelian Lie algebra with basis x, y. Classify two-dimensional $\mathfrak{g}$-modules up to isomorphism. (*Hint*: this amounts to classifying pairs of commuting matrices $X, Y \in \mathrm{Mat}_2$ up to simultaneous conjugation. Fixing Y to be a matrix in Jordan canonical form, one has to classify the elements of Mat_2 that commute with Y up to conjugation by elements of GL_2 that also commute with Y.)

Exercise 4.5.5. Recall that to a Lie algebra $\mathfrak{g}$ is associated its Lie algebra of derivations $\mathrm{Der}(\mathfrak{g})$ as in Definition 3.1.5.

(i) Show that $\mathrm{ad}(z) \in \mathrm{Der}(\mathfrak{g})$ for all $z \in \mathfrak{g}$.
(ii) Show that $\mathrm{ad}(\mathfrak{g}) = \{\mathrm{ad}(z) \mid z \in \mathfrak{g}\}$ is an ideal of $\mathrm{Der}(\mathfrak{g})$.
(iii) Show that $\mathrm{ad}(\mathfrak{l}_2) = \mathrm{Der}(\mathfrak{l}_2)$.
(iv) Find a non-abelian Lie algebra $\mathfrak{g}$ such that $\mathrm{ad}(\mathfrak{g}) \neq \mathrm{Der}(\mathfrak{g})$.

Exercise 4.5.6. Let $\mathfrak{g}$ be a Lie algebra and V a $\mathfrak{g}$-module.

(i) Show that the vector space $\mathfrak{g} \oplus V$ is a Lie algebra under the bracket

$$[(x, v), (x', v')] = ([x, x'], xv' - x'v) \quad \text{for } x, x' \in \mathfrak{g}, v, v' \in V.$$

We call this Lie algebra the *semi-direct product* $\mathfrak{g} \ltimes V$.
(ii) Use this construction to give an example of a Lie algebra $\mathfrak{h}$ such that $\mathcal{D}\mathfrak{h} = \mathfrak{h}$ and $Z(\mathfrak{h}) = \{0\}$ but $\mathfrak{h}$ is neither simple nor the direct product of simple Lie algebras.

Exercise 4.5.7. After Exercises 3.4.6 and 3.4.7, the only remaining part of the classification of three-dimensional Lie algebras is the following. Suppose that $\dim \mathfrak{g} = 3$ and $\mathcal{D}\mathfrak{g} = \mathfrak{g}$.

(i) Show that $\mathfrak{g}$ is simple.
(ii) Show that, for some $x \in \mathfrak{g}$, $\mathrm{ad}_{\mathfrak{g}}(x)$ has a nonzero eigenvalue.
(iii) Conclude that $\mathfrak{g} \cong \mathfrak{sl}_2$.

Exercise 4.5.8. Let V be a $\mathfrak{g}$-module, $\mathfrak{h}$ an ideal of $\mathfrak{g}$, and χ a character of $\mathfrak{h}$ such that $V_\chi^{\mathfrak{h}} \neq \{0\}$.

(i) Let $v \in V_\chi^{\mathfrak{h}}$ and $x \in \mathfrak{g}$. Prove by induction that, for all $i \in \mathbb{N}$ and all $y \in \mathfrak{h}$, $yx^i v - \chi(y)x^i v$ is a linear combination of $\{v, xv, \ldots, x^{i-1}v\}$.

(ii) Hence show that $\chi([y, x]) = 0$ for all $y \in \mathfrak{h}$, $x \in \mathfrak{g}$.

(iii) Deduce that $V_\chi^{\mathfrak{h}}$ is a $\mathfrak{g}$-submodule of V.

CHAPTER 5

The theory of $\mathfrak{sl}_2$-modules

In this chapter we study the representation theory of the smallest simple Lie algebra, $\mathfrak{sl}_2$. The theory as we will develop it goes back to Cartan in the 1890s, but it seems that the results were known to Lie and his student Engel even earlier.

5.1 Classification of irreducibles

Recall from Example 4.1.5 that an $\mathfrak{sl}_2$-module is a vector space V with three linear transformations e_V, h_V, f_V that satisfy the $\mathfrak{sl}_2$ bracket relations

$$[e_V, f_V] = h_V, \quad [h_V, e_V] = 2e_V, \quad [h_V, f_V] = -2f_V. \tag{5.1.1}$$

Equivalently, these equations say that, for any $v \in V$,

$$efv - fev = hv, \quad hev - ehv = 2ev, \quad hfv - fhv = -2fv. \tag{5.1.2}$$

(As was explained in the last chapter, efv is defined to mean $e(fv)$, and so on.) We will see that the three equations (5.1.1) are more restrictive than may appear; to illustrate this with an easy observation, it follows as in Proposition 4.1.8 that e_V, h_V, and f_V must have trace zero.

We first list some easy examples of $\mathfrak{sl}_2$-modules.

Example 5.1.1. The trivial module is defined by $V = \mathbb{C}$ and $e_V = h_V = f_V = 0$. ■

Example 5.1.2. The natural two-dimensional module is defined by $V = \mathbb{C}^2$ and

$$e_V = \begin{pmatrix} 0 & 1 \\ 0 & 0 \end{pmatrix}, \quad h_V = \begin{pmatrix} 1 & 0 \\ 0 & -1 \end{pmatrix}, \quad f_V = \begin{pmatrix} 0 & 0 \\ 1 & 0 \end{pmatrix}.$$

This is irreducible since there is no common eigenvector of these three matrices (or for the general reason explained in Example 4.3.10). ■

Example 5.1.3. Another example is the adjoint representation on $\mathfrak{sl}_2$ itself, for which we found the representing matrices in Example 4.1.14. This is irreducible since $\mathfrak{sl}_2$ is simple by Theorem 3.3.10. ■

Example 5.1.4. A four-dimensional example is the adjoint representation of $\mathfrak{gl}_2$ restricted to give an $\mathfrak{sl}_2$-module structure on $\mathfrak{gl}_2$. This is reducible since $\mathfrak{sl}_2$ itself is obviously an $\mathfrak{sl}_2$-submodule of $\mathfrak{gl}_2$. It is also completely reducible: the decomposition $\mathfrak{gl}_2 = \mathbb{C}1_2 \oplus \mathfrak{sl}_2$ exhibits $\mathfrak{gl}_2$ as a direct sum of irreducible submodules. ■

Example 5.1.5. Embed $\mathfrak{sl}_2$ in $\mathfrak{gl}_3$ in the top-left 2×2 block. Then $\mathfrak{gl}_3$ becomes an $\mathfrak{sl}_2$-module by restriction of the adjoint representation of $\mathfrak{gl}_3$. There is an obvious decomposition as a direct sum of $\mathfrak{sl}_2$-submodules:

$$\mathfrak{gl}_3 = \begin{pmatrix} \mathfrak{sl}_2 & 0 \\ 0 & 0 \end{pmatrix} \oplus \begin{pmatrix} \mathbb{C}1_2 & 0 \\ 0 & 0 \end{pmatrix} \oplus \begin{pmatrix} 0 & * \\ 0 & 0 \end{pmatrix} \oplus \begin{pmatrix} 0 & 0 \\ * & 0 \end{pmatrix} \oplus \begin{pmatrix} 0 & 0 \\ 0 & * \end{pmatrix},$$

where in each case the top-left entry is a 2×2 block, the top-right entry is a 2×1 block, and so on. These $\mathfrak{sl}_2$-submodules are all irreducible: the first is isomorphic to the adjoint representation of $\mathfrak{sl}_2$, and the second and fifth are the trivial module; the third and fourth are both isomorphic to the natural module $\mathbb{C}^2$, as one can see from the equations

$$\left[\begin{pmatrix} x & 0 \\ 0 & 0 \end{pmatrix}, \begin{pmatrix} 0 & v \\ 0 & 0 \end{pmatrix}\right] = \begin{pmatrix} 0 & xv \\ 0 & 0 \end{pmatrix},$$

$$\left[\begin{pmatrix} x & 0 \\ 0 & 0 \end{pmatrix}, \begin{pmatrix} 0 & 0 \\ (sv)^{\mathrm{t}} & 0 \end{pmatrix}\right] = \begin{pmatrix} 0 & 0 \\ (sxv)^{\mathrm{t}} & 0 \end{pmatrix},$$

where $x \in \mathfrak{sl}_2$, $v \in \mathbb{C}^2$ (a column vector), and $s = \left(\begin{smallmatrix} 0 & -1 \\ 1 & 0 \end{smallmatrix}\right)$. This decomposition into irreducible $\mathfrak{sl}_2$-submodules is not unique: for instance, the sum of the two trivial modules could be split up in an arbitrary way. ■

One feature that all these examples share is that the matrix representing h is diagonalizable. This is a clue that the eigenspaces of h_V are going to be a crucial feature.

Definition 5.1.6. Let V be an $\mathfrak{sl}_2$-module. For $a \in \mathbb{C}$, the a-eigenspace of h_V is written V_a and called the *weight space of weight* a, and any a-eigenvector is called a *weight vector of weight* a. The eigenvalues of h_V are called the *weights* of V; $\dim V_a$ is called the *multiplicity* a.

As with the usual conventions for eigenvectors and eigenspaces, weight vectors are nonzero by definition but weight spaces include zero (and are indeed subspaces of V). Now, if e_V and f_V commuted with h_V, they would necessarily preserve the weight spaces; since they in fact do not commute with h_V but rather satisfy the bracket relations (5.1.1), it turns out that they raise and lower weights by 2, in the following sense.

Proposition 5.1.7. *For any $a \in \mathbb{C}$, we have $eV_a \subseteq V_{a+2}$ and $fV_a \subseteq V_{a-2}$.*

Proof. If $v \in V_a$ then $hev = 2ev + ehv = 2ev + e(av) = (a+2)ev$, so $ev \in V_{a+2}$. Similarly, $hfv = -2fv + fhv = (a-2)fv$, so $fv \in V_{a-2}$. □

Definition 5.1.8. A weight vector v such that $ev = 0$ is called a *highest-weight vector.*

This term is a little unfortunate, since the weight of v need not be higher than all other weights of V; all it conveys is that you cannot create from v a weight vector of higher weight by applying e, as you might hope to do.

Proposition 5.1.9. *Any nonzero $\mathfrak{sl}_2$-module V contains a highest-weight vector of some weight.*

Proof. Since h_V has an eigenvector, there must be some nonzero weight space in V, say V_a. Since h_V has only finitely many eigenvalues, V_{a+2i} must be zero once $i \in \mathbb{N}$ is sufficiently large. Hence there is some $j \in \mathbb{N}$ such that $V_{a+2j} \neq \{0\}$, $V_{a+2j+2} = \{0\}$. Any nonzero $v \in V_{a+2j}$ is a highest-weight vector, by Proposition 5.1.7. □

Next we describe the submodule generated by a highest-weight vector; the nice thing about $\mathfrak{sl}_2$ is that simple calculations suffice to do this.

Proposition 5.1.10. *Let V be an $\mathfrak{sl}_2$-module, and suppose that $w_0 \in V$ is a highest-weight vector of weight m. Then $m \in \mathbb{N}$, and the submodule generated by w_0 has basis $w_0, w_1, \ldots, w_m$ satisfying*

$$ew_i = \begin{cases} (m-i+1)w_{i-1} & \text{if } 1 \le i \le m, \\ 0 & \text{if } i = 0, \end{cases}$$

$$fw_i = \begin{cases} (i+1)w_{i+1} & \text{if } 0 \le i \le m-1, \\ 0 & \text{if } i = m, \end{cases}$$

$$hw_i = (m-2i)w_i.$$

Proof. Define

$$w_i = \frac{1}{i!} f^i w_0 \in V_{m-2i} \quad \text{for all } i \in \mathbb{N}.$$

Since h_V has finitely many eigenvalues, $V_{m-2i} = 0$ for i sufficiently large, so there is some $j \in \mathbb{N}$ such that $w_j \neq 0$, $w_{j+1} = 0$. We know that

$$fw_i = (i+1)w_{i+1} \quad \text{and} \quad hw_i = (m-2i)w_i \qquad \text{for all } i \in \mathbb{N}.$$

We now prove by induction that $ew_i = (m - i + 1)w_{i-1}$ for all $i \geq 1$. The base case is

$$ew_1 = efw_0 = hw_0 + few_0 = mw_0 + f0 = mw_0.$$

Assuming the result to be true for some i, it follows that

$$\begin{aligned}(i+1)ew_{i+1} &= efw_i \\ &= hw_i + few_i \\ &= (m-2i)w_i + f(m-i+1)w_{i-1} \\ &= [(m-2i) + (m-i+1)i]w_i \\ &= (i+1)(m-i)w_i,\end{aligned}$$

as required. Now, from the fact that $0 = ew_{j+1} = (m - j)w_j$, we see that $j = m$, which implies that $m \in \mathbb{N}$ and that $fw_m = 0$. It only remains to note that $w_0, w_1, \ldots, w_m$ are linearly independent because they are eigenvectors for h_V corresponding to different eigenvalues. □

The action of e, f, and h on $w_0, w_1, \ldots, w_m$ can be depicted thus:

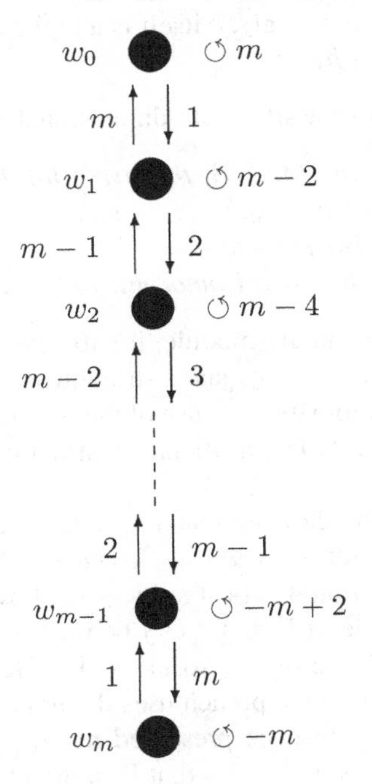

Here the up arrows represent e, the down arrows represent f, and the circular arrows represent h. The arrows are labelled by the coefficient that appears when the corresponding Lie algebra element acts on the corresponding w_i.

Definition 5.1.11. In any $\mathfrak{sl}_2$-module V, a subset $w_0, w_1, \ldots, w_m$ of nonzero vectors on which e, f, and h act by the above formulas is called a *string with highest weight* m.

Thus Proposition 5.1.10 shows that any highest-weight vector w_0 of weight m is part of a string with highest weight m, defined by

$$w_i = \frac{1}{i!} f^i w_0.$$

Example 5.1.12. In the trivial module, any nonzero vector is a string with highest (and only) weight 0. ■

Example 5.1.13. In the natural two-dimensional module of $\mathfrak{sl}_2$, the standard basis vectors v_1, v_2 form a string with highest weight 1. Of course, we can multiply both by some nonzero scalar and still have a string. ■

Example 5.1.14. In the $\mathfrak{sl}_2$-module $\mathfrak{sl}_2$, e itself is a highest-weight vector of weight 2; the string it generates is $e, -h, -f$. ■

We already know enough to classify finite-dimensional irreducibles.

Theorem 5.1.15. (1) *For any $m \in \mathbb{N}$, the string formulas do indeed define an $\mathfrak{sl}_2$-module $V(m)$ with basis $w_0, w_1, \ldots, w_m$.*

(2) *The $\mathfrak{sl}_2$-module $V(m)$ is irreducible.*

(3) *Any irreducible $\mathfrak{sl}_2$-module V is isomorphic to $V(m)$, where $m = \dim V - 1$.*

Proof. To verify that $V(m)$ is an $\mathfrak{sl}_2$-module, it suffices to show that (5.1.2) holds for all $v \in \{w_0, \ldots, w_m\}$; this involves just a straightforward check from the string formulas, much of which is automatic because of the way in which we found the formulas in the first place. A slightly less artificial construction is provided by Exercise 5.3.4.

There are two approaches to showing that $V(m)$ is irreducible; both are instructive. In the first approach we let $w = a_0 w_0 + a_1 w_1 + \cdots + a_m w_m$ be a nonzero element of $V(m)$. By Proposition 4.3.14 it suffices to show that the $\mathfrak{sl}_2$-submodule W generated by w is the whole of $V(m)$. Let j be maximal such that $a_j \neq 0$; then clearly $e^j w$ is a nonzero multiple of w_0, so $w_0 \in W$. Hence $w_i = \frac{1}{i!} f^i w_0 \in W$ for all i, so $W = V(m)$. The other approach uses the important principle that, since $h_{V(m)}$ is diagonalizable, any subspace preserved by $h_{V(m)}$ must be a direct sum of subspaces of the eigenspaces of $h_{V(m)}$, that is, a direct sum of subspaces of the weight spaces. In our case, where the weight spaces are one dimensional, this means that any nonzero $\mathfrak{sl}_2$-submodule must contain some weight space. Using powers of e and f we then see that it must contain all weight spaces and hence equals $V(m)$. Either way, (2) is proved.

For part (3), Propositions 5.1.9 and 5.1.10 show that V contains an $\mathfrak{sl}_2$-submodule isomorphic to $V(m)$ for some m; since V is irreducible this submodule must be the whole of V, so $V \cong V(m)$ and $\dim V = \dim V(m) = m + 1$. □

Remark 5.1.16. Theorem 5.1.15 implies that the irreducible $\mathfrak{sl}_2$-modules are completely classified by their dimension. Considering the situation for the abelian Lie algebras discussed in Example 4.4.6 and in the comments following Remark 4.4.8, it is noteworthy that the classification for $\mathfrak{sl}_2$ is discrete, i.e. it involves no continuous parameters. In Theorem 7.4.9 we will see that such a discrete classification holds for $\mathfrak{sl}_n$ in general.

5.2 Complete reducibility

As mentioned in the previous chapter, classifying the irreducible modules over a Lie algebra is, in general, far from being enough to classify arbitrary modules. The case of $\mathfrak{sl}_2$ is straightforward, however: every module is a direct sum of irreducibles with uniquely defined multiplicities.

Theorem 5.2.1. *Every $\mathfrak{sl}_2$-module is completely reducible.*

Proof. Let V be an $\mathfrak{sl}_2$-module, and W a nontrivial submodule of V. By induction on dimension we can assume that W is completely reducible, so that W is a direct sum of submodules isomorphic to various $V(m_i)$. We need to show that there is a submodule W' of V such that $W \oplus W' = V$. It suffices to find an irreducible submodule U such that $W \cap U = \{0\}$, because then we can repeat the argument with W replaced by $W \oplus U$, and thus build up W' piece by piece. To find U it suffices to find a highest-weight vector v_0 in V that is not in W, for then the submodule U generated by v_0 is irreducible by Theorem 5.1.15 and its submodule $W \cap U$ (which by definition is not the whole of U) must be $\{0\}$, as required.

Now consider the quotient module V/W. By Propositions 5.1.9 and 5.1.10, it contains a highest-weight vector $\overline{v_0}$ having some weight $m \in \mathbb{N}$. By definition,

$$\overline{v_0} \neq 0, \quad h\overline{v_0} = m\overline{v_0}, \quad e\overline{v_0} = 0. \tag{5.2.1}$$

Let $v_0 \in V$ be a preimage of $\overline{v_0}$; we conclude that

$$v_0 \notin W, \quad hv_0 - mv_0 \in W, \quad ev_0 \in W. \tag{5.2.2}$$

Our goal is to adjust v_0 by adding elements of W, thus preserving all the properties in (5.2.2), until it becomes a highest-weight vector of weight m. First note that W is the direct sum of its weight spaces, since every irreducible module is such a direct sum. So every element of W can be written uniquely as a sum of weight vectors. Since $h_W - m$ acts as a scalar $a - m$ on the weight space W_a, we can annihilate all the components of $hv_0 - mv_0$ with weights other than m by adding appropriate elements of those weight spaces to v_0. Thus we can ensure that $hv_0 - mv_0 \in W_m$. Now ev_0 is also in W, and we can find its decomposition into weight vectors by applying h to it. We have

$$hev_0 = ehv_0 + 2ev_0 = e(hv_0 - mv_0) + (m+2)ev_0 \tag{5.2.3}$$

so that $(h - (m+2))ev_0 \in W_{m+2}$, which implies that $ev_0 \in W_{m+2}$. From the structure of the $V(m_i)$ making up W (and remembering that $m \geq 0$) it is clear that $e : W_m \to W_{m+2}$ is surjective, so there exists $w \in W_m$ such that $ew = ev_0$. Replacing v_0 by $v_0 - w$, we can ensure that $ev_0 = 0$. Clearly this does not change $hv_0 - mv_0$.

If $w_0 := hv_0 - mv_0$ is zero we are finished, so let us assume for contradiction that $w_0 \neq 0$. Then (5.2.3) shows that $ew_0 = 0$, so w_0 is a highest-weight vector of weight m. Define $v_i = \frac{1}{i!} f^i v_0$, $w_i = \frac{1}{i!} f^i w_0$ for all $i \in \mathbb{N}$. So $w_j = 0$ for $j > m$, and $w_0, w_1, \ldots, w_m$ form a string of highest weight m. We now claim that

$$hv_i = (m - 2i)v_i + w_i \quad \text{for all } i \in \mathbb{N}. \tag{5.2.4}$$

The proof is by induction, the $i = 0$ case being the definition of w_0. Assuming the result for i, we find

$$\begin{aligned}(i+1)hv_{i+1} &= hfv_i \\ &= fhv_i - 2fv_i \\ &= (m - 2i - 2)fv_i + fw_i \\ &= (i+1)\left((m - 2i - 2)v_{i+1} + w_{i+1}\right),\end{aligned}$$

as required. So for $i > m$ we have $v_i \in V_{m-2i}$ and, as usual, we know that $V_{m-2i} = 0$ for i sufficiently large. Thus there is some $j \in \mathbb{N}$ such that $v_j \neq 0$, $v_{j+1} = 0$; by (5.2.4) and the fact that $w_m \neq 0$ we see that $j \geq m$. Finally, another induction on i proves that

$$ev_i = (m - i + 1)v_{i-1} + w_{i-1} \quad \text{for all } i \geq 1. \tag{5.2.5}$$

Setting $i = j+1$ we see that $0 = (m - j)v_j + w_j$. If $j = m$ this contradicts the fact that $w_m \neq 0$ and if $j > m$ this contradicts the fact that $v_j \neq 0$. □

Remark 5.2.2. A different proof of Theorem 5.2.1 will appear later as the $n = 2$ case of Theorem 7.5.3, which is more general.

This seems rather amazing: we started with just the three equations (5.1.1) and the conclusion is that with respect to a suitable basis, namely, the union of the strings in the various irreducible summands, e_V, h_V, and f_V are given by the explicit formulas of Proposition 5.1.10.

Corollary 5.2.3. *Let V be any $\mathfrak{sl}_2$-module.*

(1) *The linear transformation h_V is diagonalizable with integer eigenvalues. In other words, V is the direct sum of its weight spaces and all its weights are integers.*

(2) *The weight multiplicities satisfy*

$$\dim V_k = \dim V_{-k},$$
$$\dim V_0 \geq \dim V_2 \geq \dim V_4 \geq \cdots,$$

and

$$\dim V_1 \geq \dim V_3 \geq \dim V_5 \geq \cdots.$$

(3) *When V is written as a direct sum of irreducible submodules, the number of summands that are isomorphic to $V(m)$ is* $\dim V_m - \dim V_{m+2}$. *Hence the isomorphism class of V is determined by its weight multiplicities.*

Proof. Parts (1) and (2) are both consequences of the fact that V can be written as a direct sum of the $V(m)$ for various m, where $V(m)$ has weights $m, m-2, \ldots, -m+2, -m$, all of multiplicity 1. This also implies that $\dim V_k$ equals the number of direct summands isomorphic to $V(m)$ where $m \geq |k|$ and $m - k$ is even. Part (3) follows. □

The impact of what we have proved goes beyond $\mathfrak{sl}_2$, since in many other Lie algebras $\mathfrak{g}$ there exist *$\mathfrak{sl}_2$-triples*, i.e. triples of elements e, h, f satisfying the $\mathfrak{sl}_2$ commutation relations. If V is a $\mathfrak{g}$-module, we can view V as an $\mathfrak{sl}_2$-module by restricting to such a triple; we can then apply Corollary 5.2.3 to obtain constraints on the eigenspaces of h_V.

5.3 Exercises

Exercise 5.3.1. Suppose that an $\mathfrak{sl}_2$-module V has the following weights (each written as many times as its multiplicity in V):

$$5, 4, 3, 3, 2, 1, 1, 0, 0, 0, -1, -1, -2, -3, -3, -4, -5.$$

The $\mathfrak{sl}_2$-module V is a direct sum of how many irreducible submodules, and of which types?

Exercise 5.3.2. Knowing the results of this chapter, how would you carry out the calculation in Exercise 2.3.5(iii)?

Exercise 5.3.3. Suppose that we have an $\mathfrak{sl}_2$-action on $\mathbb{C}^n$ in which the matrix representing e is the Jordan block matrix with ones above the diagonal and all other entries zero and the matrix representing h is diagonal. Find the matrix representing f.

Exercise 5.3.4. For $m \in \mathbb{N}$, let S^m be the vector space of homogeneous polynomials p of degree m in two indeterminates t and u.

(i) Show that there is an $\mathfrak{sl}_2$-action on S^m defined by

$$ep = t\frac{\partial p}{\partial u}, \qquad hp = t\frac{\partial p}{\partial t} - u\frac{\partial p}{\partial u}, \qquad fp = u\frac{\partial p}{\partial t}.$$

(ii) Show that the basis

$$t^m, \quad mt^{m-1}u, \quad \binom{m}{2}t^{m-2}u^2, \quad \cdots, \quad mtu^{m-1}, \quad u^m$$

of S^m is a string with highest weight m. Thus $S^m \cong V(m)$.

Exercise 5.3.5. For $n \in \mathbb{N}$, let V be a vector space with basis v_I indexed by all the subsets I of the set $\{1, \ldots, n\}$. Thus $\dim V = 2^n$.

(i) Show that there is an $\mathfrak{sl}_2$-action on V defined by

$$ev_I = \sum_{\substack{J \supset I \\ |J \setminus I| = 1}} v_J, \qquad hv_I = (2|I| - n)v_I, \qquad fv_I = \sum_{\substack{K \subset I \\ |I \setminus K| = 1}} v_K.$$

(ii) Determine the multiplicity of each irreducible $\mathfrak{sl}_2$-module $V(m)$ in V.

Exercise 5.3.6. Parts of the $\mathfrak{sl}_2$ theory have analogues for an $\mathfrak{l}_2$-module V, in which s and t play roles similar to those of h and e, respectively. As we saw in Exercise 4.5.4, it is not always true that s_V is diagonalizable but V is certainly the direct sum of the generalized eigenspaces $V_a^{\text{gen}} = \{v \in V \mid (s_V - a)^k(v) = 0 \text{ for some } k \in \mathbb{N}\}$ as a runs over $\mathbb{C}$.

(i) Show that, for all $a \in \mathbb{C}$, $tV_a^{\text{gen}} \subseteq V_{a+1}^{\text{gen}}$.
(ii) Deduce that t_V is nilpotent.
(iii) Show that V contains a one-dimensional submodule. Hence the only irreducible $\mathfrak{l}_2$-modules are those that are one-dimensional.

Exercise 5.3.7. Let V be an $\mathfrak{sl}_2$-module.

(i) For any $m \in \mathbb{N}$, let $V_{[m]}$ be the $\mathfrak{sl}_2$-submodule of V generated by all the highest-weight vectors of weight m in V. (Thus $V_{[m]} = \{0\}$ if V does not contain any highest-weight vectors of weight m.) Show that $V = \bigoplus_{m \in \mathbb{N}} V_{[m]}$.

(ii) Deduce that there is a unique linear transformation τ of V such that for every string $w_0, w_1, \dots, w_m$ we have $\tau(w_i) = w_{m-i}$ for all i.

(iii) Show that $\tau(ev) = f\tau(v)$, $\tau(hv) = -h\tau(v)$, and $\tau(fv) = e\tau(v)$ for all $v \in V$.

Exercise 5.3.8. In this exercise, breaking our standing convention, we consider some infinite-dimensional modules, to which the results of this chapter do not apply. Let $a, b \in \mathbb{C}$, and define a vector space W with basis $\dots, w_{-2}, w_{-1}, w_0, w_1, w_2, \dots$. Define linear transformations h_W, f_W of W by the rules $h_W(w_i) = (a + 2i)w_i$ and $f_W(w_i) = w_{i-1}$, for all $i \in \mathbb{Z}$.

(i) Show that, subject to the constraint $e_W(w_0) = bw_1$, there is a unique way to define e_W so as to complete an $\mathfrak{sl}_2$-action on W.

(ii) Show that any nonzero $\mathfrak{sl}_2$-submodule of W contains w_i for some i.

(iii) Show that W is reducible if and only if $b = ja + j(j+1)$ for some $j \in \mathbb{Z}$.

CHAPTER 6

General theory of modules

In this chapter we return to the theory of modules over an arbitrary Lie algebra $\mathfrak{g}$. There is a great deal that could be said, even in this generality, but we will concentrate on the concepts needed for our study of integral $\mathfrak{gl}_n$-modules in the next chapter.

6.1 Duals and tensor products

A feature that the theory of modules over Lie algebras shares with the representation theory of groups is that there is a natural operation of duality. Recall that, for any vector space V, the dual space V^* consists of all the linear functions $f : V \to \mathbb{C}$.

Proposition 6.1.1. *Let V be any $\mathfrak{g}$-module.*

(1) *We can define a $\mathfrak{g}$-action on V^* by the rule*

$$(xf)(v) = -f(xv) \quad \textit{for all } x \in \mathfrak{g},\ f \in V^*,\ v \in V.$$

(2) *The natural identification of $(V^*)^*$ with V is a $\mathfrak{g}$-module isomorphism.*
(3) *The $\mathfrak{g}$-module V is irreducible if and only if V^* is irreducible.*

Proof. In (1), all that requires checking is that $[x, y]f = xyf - yxf$ for all $x, y \in \mathfrak{g}$ and $f \in V^*$. For any $v \in V$, we have $(xyf)(v) = -(yf)(xv) = f(yxv)$, so

$$(xyf - yxf)(v) = f(yxv - xyv) = f([y, x]v) = -f([x, y]v) = ([x, y]f)(v),$$

as required. In (2), the identification referred to is that identifying $v \in V$ with the linear function $V^* \to \mathbb{C} : f \mapsto f(v)$, and it is immediate from the definition that this respects the $\mathfrak{g}$-action. Because of (2) we need only prove one direction of (3). Suppose that V^* is irreducible, and let W be a submodule of V. From the definition it follows easily that the 'perpendicular subspace' $W^\perp = \{f \in V^* \mid f(W) = 0\}$ is a submodule of V^*, so either $W^\perp = \{0\}$ or $W^\perp = V^*$. Since $\dim W^\perp = \dim V - \dim W$, this implies that either $W = V$ or $W = \{0\}$. Hence V is irreducible. □

Recall from Example 1.2.1 that if V has a basis $v_1, \ldots, v_n$ then the dual basis $v_1^*, \ldots, v_n^*$ of V^* is defined by $v_i^*(v_j) = \delta_{ij}$. The basis of $(V^*)^*$ that is dual to the basis $v_1^*, \ldots, v_n^*$ of V^* is identified with the original basis of V, so we can simply say that the two bases of V and V^* are dual to each other.

Proposition 6.1.2. *Let V be a $\mathfrak{g}$-module, $x \in \mathfrak{g}$. With respect to bases of V and V^* that are dual to each other, the matrix of x_{V^*} is the negative transpose of that of x_V.*

Proof. The (i, j) entry of the matrix of x_V is $v_i^*(xv_j)$, and the (j, i) entry of the matrix of x_{V^*} is $(xv_i^*)(v_j) = -v_i^*(xv_j)$. □

Example 6.1.3. The dual of the one-dimensional module $\mathbb{C}_\chi$ is isomorphic to $\mathbb{C}_{-\chi}$, for any character χ of $\mathfrak{g}$. ■

Example 6.1.4. By Proposition 6.1.1, we know that the $\mathfrak{sl}_2$-module $V(m)^*$ is irreducible, so it must be isomorphic to $V(m)$. Using Proposition 6.1.2 and the string formulas of Proposition 5.1.10, we can write down the action of e, f, h on the basis $w_0^*, w_1^*, \ldots, w_m^*$ dual to the string basis of $V(m)$:

$$\begin{aligned} ew_i^* &= -(m-i)w_{i+1}^* \quad \text{(which equals 0 if } i = m), \\ fw_i^* &= -iw_{i-1}^* \quad \text{(which equals 0 if } i = 0), \\ hw_i^* &= -(m-2i)w_i^*. \end{aligned}$$

In particular, w_m^* is a highest-weight vector, and the string it generates is

$$w_m^*, -mw_{m-1}^*, \binom{m}{2} w_{m-2}^*, \ldots, (-1)^m w_0^*.$$

So we have an explicit isomorphism $V(m) \xrightarrow{\sim} V(m)^*$ given by $w_i \mapsto (-1)^i \binom{m}{i} w_{m-i}^*$. ■

As well as taking the dual of a $\mathfrak{g}$-module, we can take the tensor product of two $\mathfrak{g}$-modules using the tensor product of vector spaces as in Section 1.2. The definition of the $\mathfrak{g}$-action is given by the following result.

Proposition 6.1.5. *If V and W are $\mathfrak{g}$-modules, there is a unique $\mathfrak{g}$-action on $V \otimes W$ for which*

$$x(v \otimes w) = xv \otimes w + v \otimes xw \quad \textit{for all } x \in \mathfrak{g},\ v \in V,\ w \in W. \tag{6.1.1}$$

Proof. Let $v_1, \ldots, v_n$ be a basis of V and $w_1, \ldots, w_m$ be a basis of W; then the elements $v_i \otimes w_j$, for $1 \leq i \leq n$ and $1 \leq j \leq m$, form a basis of $V \otimes W$. We want to define the action $x_{V \otimes W}$, for $x \in \mathfrak{g}$, in such a way that it satisfies the rule (6.1.1). We can specify that this rule holds for the basis elements, i.e. $x(v_i \otimes w_j) = xv_i \otimes w_j + v_i \otimes xw_j$; this uniquely defines a linear transformation $x_{V \otimes W}$. Then, since both sides of (6.1.1) are bilinear functions of v and w, the fact that (6.1.1) holds when v belongs to a particular basis of V and w belongs to a particular basis of W implies that it holds always. It is clear that $x \mapsto x_{V \otimes W}$ is linear, so all that remains is to prove that $[x, y]_{V \otimes W} = [x_{V \otimes W}, y_{V \otimes W}]$ for all $x, y \in \mathfrak{g}$. Since the pure tensors span $V \otimes W$, we need only show that these linear transformations agree on a pure tensor $v \otimes w$. We have

$$\begin{aligned} xy(v \otimes w) &= x(yv \otimes w + v \otimes yw) \\ &= xyv \otimes w + yv \otimes xw + xv \otimes yw + v \otimes xyw, \end{aligned}$$

and

$$yx(v \otimes w) = yxv \otimes w + xv \otimes yw + yv \otimes xw + v \otimes yxw.$$

Taking the difference gives

$$xy(v \otimes w) - yx(v \otimes w) = [x, y]v \otimes w + v \otimes [x, y]w,$$

which proves the claim. □

Remark 6.1.6. The fact that the representation of $\mathfrak{g}$ on $V \otimes W$ is similar to the product rule for differentiation is no coincidence: this representation can be obtained by differentiating the representation of the corresponding Lie group, in the sense discussed in the first chapter.

Example 6.1.7. Recall the representation of $\mathfrak{l}_2$ on $\mathbb{C}^2$ in which s has representing matrix $\left(\begin{smallmatrix} 1 & 0 \\ 0 & 0 \end{smallmatrix}\right)$ and t has representing matrix $\left(\begin{smallmatrix} 0 & 1 \\ 0 & 0 \end{smallmatrix}\right)$. In the action of $\mathfrak{l}_2$ on $\mathbb{C}^2 \otimes \mathbb{C}^2$ we have, for instance,

$$t(v_1 \otimes v_2) = tv_1 \otimes v_2 + v_1 \otimes tv_2 = 0 + v_1 \otimes v_1 = v_1 \otimes v_1.$$

In this way we calculate that the representing matrices with respect to the basis $v_1 \otimes v_1, v_1 \otimes v_2, v_2 \otimes v_1, v_2 \otimes v_2$ are

$$s_{\mathbb{C}^2 \otimes \mathbb{C}^2} : \begin{pmatrix} 2 & 0 & 0 & 0 \\ 0 & 1 & 0 & 0 \\ 0 & 0 & 1 & 0 \\ 0 & 0 & 0 & 0 \end{pmatrix}, \qquad t_{\mathbb{C}^2 \otimes \mathbb{C}^2} : \begin{pmatrix} 0 & 1 & 1 & 0 \\ 0 & 0 & 0 & 1 \\ 0 & 0 & 0 & 1 \\ 0 & 0 & 0 & 0 \end{pmatrix}.$$ ■

Example 6.1.8. Taking tensor products of one-dimensional modules adds their characters: $\mathbb{C}_\chi \otimes \mathbb{C}_{\chi'} \cong \mathbb{C}_{\chi+\chi'}$. ■

Example 6.1.9. There is a canonical vector space isomorphism $\mathbb{C} \otimes V \xrightarrow{\sim} V$ that maps $1 \otimes v$ to v for all $v \in V$. If V is a $\mathfrak{g}$-module and we give $\mathbb{C}$ the trivial $\mathfrak{g}$-module structure then this is clearly an isomorphism of $\mathfrak{g}$-modules. If we choose a nonzero character χ of $\mathfrak{g}$ then $\mathbb{C}_\chi \otimes V$ can be identified with V as a vector space but it has a different $\mathfrak{g}$-module structure, in which the scalar $\chi(x)1_V$ is added to the action of each $x \in \mathfrak{g}$. Note that the submodules of V are the same (under the identification of vector spaces) as the submodules of $\mathbb{C}_\chi \otimes V$. In particular, $\mathbb{C}_\chi \otimes V$ is irreducible if and only if V is irreducible. ■

Example 6.1.10. To illustrate the tensor product construction further, we now find the weight multiplicities, and hence the decomposition into irreducibles, of the $\mathfrak{sl}_2$-module $V(m) \otimes V(n)$. Since $V(n) \otimes V(m)$ is isomorphic to $V(m) \otimes V(n)$ via the map that swaps the tensor factors, we can assume without loss of generality that $m \geq n$. Write $v_0, v_1, \ldots, v_m$ for a string basis of $V(m)$ and $w_0, w_1, \ldots, w_n$ for a string basis of $V(n)$, so that $V(m) \otimes V(n)$ has basis $v_i \otimes w_j$, $0 \leq i \leq m$, $0 \leq j \leq n$. We calculate

$$\begin{aligned} h(v_i \otimes w_j) &= hv_i \otimes w_j + v_i \otimes hw_j, \\ &= (m - 2i)v_i \otimes w_j + v_i \otimes (n - 2j)w_j, \\ &= (m + n - 2i - 2j)(v_i \otimes w_j), \end{aligned}$$

so $v_i \otimes w_j$ is a weight vector of weight $m + n - 2(i + j)$. It follows that the weights of $V(m) \otimes V(n)$ are the integers between $-(m + n)$ and $m + n$ of the same parity as $m + n$; and, for any such integer a, the weight multiplicity $\dim(V(m) \otimes V(n))_a$ is the number of pairs (i, j), $0 \leq i \leq m$, $0 \leq j \leq n$ such that $i + j = \frac{1}{2}(m + n - a)$. It is easy to see that this number is

$$\begin{cases} \frac{1}{2}(m + n - a) + 1 & \text{if } a \geq m - n, \\ n + 1 & \text{if } n - m < a < m - n, \\ \frac{1}{2}(m + n + a) + 1 & \text{if } a \leq n - m. \end{cases} \tag{6.1.2}$$

Applying Corollary 5.2.3(3), we deduce that

$$V(m) \otimes V(n) \cong V(m + n) \oplus V(m + n - 2) \oplus \cdots \oplus V(m - n), \tag{6.1.3}$$

where the highest weights on the right decrease by 2 at each step. To take a specific case, the basis vectors of $V(3) \otimes V(2)$ may be represented as dots in a 4×3 grid:

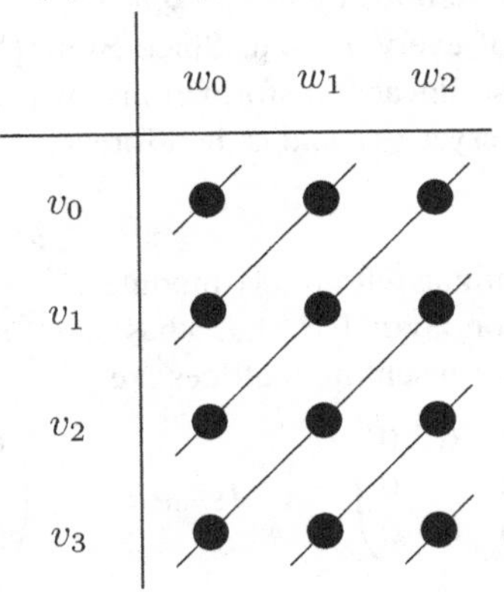

The basis vectors on each diagonal have the same weight, so we conclude that the weights are ± 5 (multiplicity 1), ± 3 (multiplicity 2), and ± 1 (multiplicity 3). Thus $V(3) \otimes V(2) \cong V(5) \oplus V(3) \oplus V(1)$. ■

In general, we have a $\mathfrak{g}$-action on the k-fold tensor product $V^{(1)} \otimes \cdots \otimes V^{(k)}$ of $\mathfrak{g}$-modules, which is still given by the product rule:

$$\begin{aligned} x(v^{(1)} \otimes v^{(2)} \otimes \cdots \otimes v^{(k)}) = xv^{(1)} &\otimes v^{(2)} \otimes \cdots \otimes v^{(k)} \\ &+ v^{(1)} \otimes xv^{(2)} \otimes \cdots \otimes v^{(k)} \\ &+ \cdots + v^{(1)} \otimes \cdots \otimes v^{(k-1)} \otimes xv^{(k)}, \end{aligned} \tag{6.1.4}$$

for all $v^{(i)} \in V^{(i)}$. There is no need for a separate verification that this does give a $\mathfrak{g}$-module, because one can build up the k-fold tensor product by repeatedly taking two-fold tensor products as above. Since the action is symmetric in the different factors of the tensor product, changing the order gives an isomorphic $\mathfrak{g}$-module. More precisely, for any permutation $\sigma \in S_k$,

$$V^{(\sigma(1))} \otimes \cdots \otimes V^{(\sigma(k))} \cong V^{(1)} \otimes \cdots \otimes V^{(k)}, \tag{6.1.5}$$

via the isomorphism defined by

$$v^{(\sigma(1))} \otimes \cdots \otimes v^{(\sigma(k))} \mapsto v^{(1)} \otimes \cdots \otimes v^{(k)}.$$

If we take the $\mathfrak{g}$-modules $V^{(1)}, \ldots, V^{(k)}$ to be the same as each other, we get the concept of the kth tensor power $V^{\otimes k} = V \otimes \cdots \otimes V$ of a $\mathfrak{g}$-module. Another consequence of the symmetry in the definition of the $\mathfrak{g}$-action is that, as in the first chapter, the symmetric and alternating tensors are preserved.

Proposition 6.1.11. *If V is a $\mathfrak{g}$-module,* $\mathrm{Sym}^k(V)$ *and* $\mathrm{Alt}^k(V)$ *are $\mathfrak{g}$-submodules of $V^{\otimes k}$.*

Proof. In the case $V^{(1)} = \cdots = V^{(k)} = V$, the isomorphism (6.1.5) says that the linear transformation of $V^{\otimes k}$ defined by $v^{(\sigma(1))} \otimes \cdots \otimes v^{(\sigma(k))} \mapsto v^{(1)} \otimes \cdots \otimes v^{(k)}$ commutes with the action of every $x \in \mathfrak{g}$. Since $\mathrm{Sym}^k(V)$ is the intersection of the 1-eigenspaces of all these linear transformations, as σ runs over S_k, it must be preserved by the action of every $x \in \mathfrak{g}$ and is therefore a $\mathfrak{g}$-submodule. The argument for $\mathrm{Alt}^k(V)$ is analogous. □

Example 6.1.12. Let us continue with the $\mathfrak{l}_2$-module $\mathbb{C}^2 \otimes \mathbb{C}^2$ as in Example 6.1.7. The space of symmetric tensors $\mathrm{Sym}^2(\mathbb{C}^2)$ has a basis $v_1 \otimes v_1, v_1 \otimes v_2 + v_2 \otimes v_1, v_2 \otimes v_2$; relative to this basis the representing matrices are

$$s_{\mathrm{Sym}^2(\mathbb{C}^2)} : \begin{pmatrix} 2 & 0 & 0 \\ 0 & 1 & 0 \\ 0 & 0 & 0 \end{pmatrix}, \qquad t_{\mathrm{Sym}^2(\mathbb{C}^2)} : \begin{pmatrix} 0 & 2 & 0 \\ 0 & 0 & 1 \\ 0 & 0 & 0 \end{pmatrix}.$$

The space of alternating tensors $\mathrm{Alt}^2(\mathbb{C}^2)$ is one-dimensional, spanned by the element $v_1 \otimes v_2 - v_2 \otimes v_1$; $\mathfrak{l}_2$ acts via the character that sends s to 1 and t to 0 (the latter of which could go without saying, because every character of $\mathfrak{l}_2$ must send $t = [s,t]$ to 0). ■

6.2 Hom-spaces and bilinear forms

Combining the operations of taking a dual and forming a tensor product, we can obtain various other natural operations. For instance, if V and W are vector spaces then the set of linear maps $\varphi : V \to W$ is a vector space that is usually written as $\mathrm{Hom}(V, W)$ ('Hom' is short for 'homomorphism') on the grounds that linear maps are the homomorphisms in the category of modules over $\mathbb{C}$. There is a canonical vector space isomorphism $V^* \otimes W \xrightarrow{\sim} \mathrm{Hom}(V, W)$ under which the pure tensor $f \otimes w$, for $f \in V^*$ and $w \in W$, corresponds to the linear map $V \to W : v \mapsto f(v)w$. In terms of chosen bases $v_1, \ldots, v_n$ of V and $w_1, \ldots, w_m$ of W, the basis element $v_i^* \otimes w_j$ of $V^* \otimes W$ corresponds to the linear map whose matrix has a 1 in the (j, i) position and zeroes elsewhere.

If V and W are $\mathfrak{g}$-modules then we know how to define a $\mathfrak{g}$-action on $V^* \otimes W$ and we can carry this across the isomorphism to $\mathrm{Hom}(V, W)$.

Proposition 6.2.1. *The $\mathfrak{g}$-action on $\mathrm{Hom}(V, W)$ resulting from its identification with $V^* \otimes W$ is given by the following rule:*

$$(x\varphi)(v) = x\varphi(v) - \varphi(xv) \quad \textit{for all } x \in \mathfrak{g},\ \varphi \in \mathrm{Hom}(V, W),\ v \in V.$$

Proof. Choosing bases of V and W as above, it suffices to verify the rule in the case where φ is the linear map that corresponds to $v_i^* \otimes w_j$ in $V^* \otimes W$. Then $x\varphi$ corresponds to $xv_i^* \otimes w_j + v_i^* \otimes xw_j$, so that

$$(x\varphi)(v) = (xv_i^*)(v)w_j + v_i^*(v)xw_j = -v_i^*(xv)w_j + xv_i^*(v)w_j = -\varphi(xv) + x\varphi(v)$$

as required. □

We can now rephrase Definition 4.2.1: a $\mathfrak{g}$-module homomorphism from V to W is a linear map $\varphi : V \to W$ such that $x\varphi = 0$ for all $x \in \mathfrak{g}$, i.e. φ is an invariant vector for $\mathfrak{g}$ in $\mathrm{Hom}(V, W)$ in the sense of Definition 4.3.5. So $\mathrm{Hom}_{\mathfrak{g}}(V, W)$ is the same as $\mathrm{Hom}(V, W)^{\mathfrak{g}}$.

Example 6.2.2. Suppose that $\mathfrak{g}$ is a subalgebra of $\mathfrak{gl}(V)$, and take $W = V$. Then the $\mathfrak{g}$-action on $\mathrm{Hom}(V, V) = \mathfrak{gl}(V)$ defined in Proposition 6.2.1 is just the restriction of the adjoint representation of $\mathfrak{gl}(V)$ to $\mathfrak{g}$. ■

Example 6.2.3. For nonnegative integers $m \geq n$, the isomorphism (6.1.3) and the fact that $V(m)^* \cong V(m)$ jointly imply that we have an $\mathfrak{sl}_2$-module isomorphism

$$\mathrm{Hom}(V(m), V(n)) \cong V(m+n) \oplus V(m+n-2) \oplus \cdots \oplus V(m-n). \quad (6.2.1)$$

Note that, if $m = n$, the last summand on the right-hand side of (6.2.1) is $V(0)$; this means that there is a one-dimensional subspace of invariant vectors for $\mathfrak{sl}_2$ in $\mathrm{Hom}(V(m), V(m))$, i.e. that $\mathrm{Hom}_{\mathfrak{sl}_2}(V(m), V(m))$ is one-dimensional. This is clear anyway from the string basis, because an $\mathfrak{sl}_2$-module endomorphism of $V(m)$ must send the highest-weight vector w_0 to another element aw_0 of the weight space $V(m)_m$, and hence it must be scalar multiplication by a since w_0 generates $V(m)$. If $m > n$ then $V(0)$ does not appear as a submodule of $\mathrm{Hom}(V(m), V(n))$, which means that there are no nonzero $\mathfrak{sl}_2$-module homomorphisms $V(m) \to V(n)$; this too can be seen from the string basis, because there is no weight vector of weight m in $V(n)$ to which to send the highest-weight vector of $V(m)$. ∎

Given any vector space V, we can consider the vector space $\mathrm{Bil}(V)$ of bilinear forms $B : V \times V \to \mathbb{C}$. There is a canonical vector space isomorphism $\mathrm{Bil}(V) \xrightarrow{\sim} \mathrm{Hom}(V, V^*)$ that sends the bilinear form B to the linear map $V \to V^* : v \mapsto B(v, \cdot)$, where $B(v, \cdot)$ denotes the linear function $V \to \mathbb{C} : v' \mapsto B(v, v')$. If V is a $\mathfrak{g}$-module then we can use this isomorphism to define a $\mathfrak{g}$-action on $\mathrm{Bil}(V)$, which is easily seen to be given by the formula

$$(xB)(v, v') = -B(xv, v') - B(v, xv'). \quad (6.2.2)$$

Definition 6.2.4. A bilinear form B on the $\mathfrak{g}$-module V is said to be $\mathfrak{g}$-*invariant* if it is an invariant vector for $\mathfrak{g}$ in $\mathrm{Bil}(V)$, i.e.

$$B(xv, v') + B(v, xv') = 0 \quad \text{for all } x \in \mathfrak{g},\ v, v' \in V.$$

Equivalently, B is $\mathfrak{g}$-invariant if the corresponding map $V \to V^* : v \mapsto B(v, \cdot)$ is a $\mathfrak{g}$-module homomorphism.

A bilinear form $B \in \mathrm{Bil}(V)$ is said to be *nondegenerate* if the corresponding map $V \to V^*$ is an isomorphism. Since $\dim V = \dim V^*$, this is equivalent to requiring that the map $V \to V^*$ has zero kernel, i.e. that the only $v \in V$ that satisfies $B(v, v') = 0$ for all $v' \in V$ is $v = 0$.

Example 6.2.5. The 'dot product' bilinear form on $\mathbb{C}^n$ given by $(v, v') \mapsto v^{\mathsf{t}} v'$ is $\mathfrak{so}_n$-invariant since, for any $x \in \mathfrak{so}_n$ and $v, v' \in \mathbb{C}^n$,

$$(xv)^{\mathsf{t}} v' + v^{\mathsf{t}} x v' = v^{\mathsf{t}}(x^{\mathsf{t}} + x) v' = 0.$$

This form is nondegenerate, so it gives an isomorphism $\mathbb{C}^n \cong (\mathbb{C}^n)^*$ of $\mathfrak{so}_n$-modules. This is not an isomorphism of $\mathfrak{gl}_n$-modules, because the bilinear form is not $\mathfrak{gl}_n$-invariant. In fact, $\mathfrak{so}_n$ is the largest subalgebra of $\mathfrak{gl}_n$ for which the bilinear form is invariant, by the following result. ∎

Proposition 6.2.6. *Let B be a symmetric nondegenerate bilinear form on the $\mathfrak{g}$-module V.*

(1) *There is a basis $v_1, \dots, v_n$ of V such that $B(v_i, v_j) = \delta_{ij}$.*
(2) *The form B is $\mathfrak{g}$-invariant if and only if, with respect to a basis as in (1), the matrices representing $\mathfrak{g}$ lie in $\mathfrak{so}_n$.*

Proof. Part (1) is proved by induction on $n = \dim V$, the zero-dimensional case being vacuously true. Assume that $n \geq 1$. Since B is symmetric, we have $B(v, w) = \frac{1}{2}(B(v + w, v + w) - B(v, v) - B(w, w))$ for all $v, w \in V$. So the fact that B is nonzero implies that there is some $v \in V$ such that $B(v, v) \neq 0$. Let $v_n = v/\sqrt{B(v, v)}$, where $\sqrt{B(v, v)}$ denotes either square root of $B(v, v)$ in $\mathbb{C}$; thus $B(v_n, v_n) = 1$. Since B is nondegenerate, the perpendicular subspace $W = \{v \in V \mid B(v, v_n) = 0\}$ is $(n - 1)$-dimensional. Since $v_n \notin W$ the restriction of B to W is still nondegenerate, so W has a basis $v_1, \dots, v_{n-1}$ of the required form by the inductive hypothesis. We can then append v_n to this basis of W to form the required basis of V.

By bilinearity, B is $\mathfrak{g}$-invariant if and only if $B(xv_i, v_j) = -B(v_i, xv_j)$ for all i and j and $x \in \mathfrak{g}$. By the assumed property of $v_1, \dots, v_n$, this equation can be rewritten as $v_j^*(xv_i) = -v_i^*(xv_j)$, which says that the (j, i) entry of the matrix representing x is the negative of the (i, j) entry. Part (2) follows. □

Example 6.2.7. The $\mathfrak{sl}_2$-module isomorphism $V(m) \xrightarrow{\sim} V(m)^*$ found in Example 6.1.4 implies that the bilinear form B on $V(m)$ defined by $B(w_i, w_j) = (-1)^i \binom{m}{i} \delta_{i+j,m}$ is nondegenerate and $\mathfrak{sl}_2$-invariant. Note that this form is symmetric if m is even and skew-symmetric (meaning that $B(v, v') = -B(v', v)$) if m is odd. In particular, with respect to a basis of $V(2)$ as in Proposition 6.2.6, the representation of $\mathfrak{sl}_2$ gives rise to an injective homomorphism $\mathfrak{sl}_2 \to \mathfrak{so}_3$. This must be an isomorphism, since both Lie algebras are three-dimensional; thus we recover the result of Exercise 2.3.5(ii). See Exercises 6.5.7 and 6.5.8 for other applications of the same idea. ■

Composing the canonical vector space isomorphisms

$$V^* \otimes V^* \xrightarrow{\sim} \mathrm{Hom}(V, V^*) \text{ and } \mathrm{Hom}(V, V^*) \xrightarrow{\sim} \mathrm{Bil}(V),$$

we get an isomorphism $V^* \otimes V^* \xrightarrow{\sim} \mathrm{Bil}(V)$ under which the pure tensor $f \otimes f'$ is mapped to the bilinear form $(v, v') \mapsto f(v)f'(v')$. The $\mathfrak{g}$-module endomorphism $\sigma : V^* \otimes V^* \to V^* \otimes V^*$ that swaps the two tensor factors corresponds to the transposing map $B \mapsto B^{\mathrm{t}}$, where B^{t} is defined by an interchange of the inputs of B: $B^{\mathrm{t}}(v, v') = B(v', v)$. The eigenspaces $\mathrm{Sym}^2(V^*)$ and $\mathrm{Alt}^2(V^*)$ of σ correspond to the submodules of $\mathrm{Bil}(V)$ consisting of symmetric and skew-symmetric bilinear forms, respectively.

6.3 Schur's lemma and the Killing form

We have already used several times the principle that if two linear transformations x and y of a vector space commute with each other then y must preserve every eigenspace of x. (The argument was spelled out in the proof of Proposition 3.2.10.) One of the most fundamental applications of this principle is the following result, which generalizes what we saw in Example 6.2.3.

Proposition 6.3.1. *[Schur's lemma] Let V, W be irreducible $\mathfrak{g}$-modules.*

(1) *If $V \ncong W$ then $\mathrm{Hom}_{\mathfrak{g}}(V, W) = \{0\}$, i.e. there are no nonzero $\mathfrak{g}$-module homomorphisms from V to W.*
(2) $\mathrm{Hom}_{\mathfrak{g}}(V, V) = \mathbb{C}1_V$, *i.e. the only $\mathfrak{g}$-module endomorphisms of V are the scalar multiplications.*
(3) *If $V \cong W$ then $\dim \mathrm{Hom}_{\mathfrak{g}}(V, W) = 1$, i.e. any two nonzero $\mathfrak{g}$-module homomorphisms $V \to W$ are scalar multiples of each other and must be isomorphisms.*

Proof. Let $\varphi : V \to W$ be a $\mathfrak{g}$-module homomorphism. We know that $\ker(\varphi)$ is a submodule of V and $\mathrm{im}(\varphi)$ is a submodule of W, but V and W, being irreducible, each have only the trivial submodules. If φ is nonzero then $\ker(\varphi)$ is not the whole of V, so it must be $\{0\}$ (i.e. φ is injective), and $\mathrm{im}(\varphi)$ is not $\{0\}$ so it must be W (i.e. φ is surjective). Thus every nonzero $\mathfrak{g}$-module homomorphism from V to W is an isomorphism, which proves (1).

Now, if $W = V$ then φ is a linear transformation of V and must have at least one eigenvalue, say a; let V_a^φ be the eigenspace. By assumption, every x_V for $x \in \mathfrak{g}$ commutes with φ and therefore preserves V_a^φ. So V_a^φ is a nonzero submodule of V and must therefore be the whole of V. This means that φ is scalar multiplication by a, proving (2).

Finally, suppose that $\varphi, \psi : V \to W$ are two nonzero $\mathfrak{g}$-module homomorphisms; by what was proved above they must be isomorphisms, so we can compose φ with the inverse of ψ to get a $\mathfrak{g}$-module isomorphism $\psi^{-1}\varphi : V \to V$. By part (2) this must be scalar multiplication by some $a \in \mathbb{C}^\times$, which implies that $\varphi = a\psi$. So part (3) is proved. □

Corollary 6.3.2. *If V is an irreducible $\mathfrak{g}$-module then every element of $Z(\mathfrak{g})$ acts on V as a scalar multiplication.*

Proof. If $z \in Z(\mathfrak{g})$ then $[x_V, z_V] = [x, z]_V = 0$ for all $x \in \mathfrak{g}$, so z_V is a $\mathfrak{g}$-module endomorphism of V. The result follows from part (2) of Schur's lemma (Proposition 6.3.1). □

This has a consequence that will be important in the next chapter.

Proposition 6.3.3. *Let V be an irreducible $\mathfrak{gl}_n$-module. Then 1_n acts on V as a scalar multiplication, and restricting the $\mathfrak{gl}_n$-action on V to $\mathfrak{sl}_n$ gives an irreducible $\mathfrak{sl}_n$-module. The resulting map*

$$\begin{aligned}&\{\text{isomorphism classes of irreducible } \mathfrak{gl}_n\text{-modules}\}\\ &\qquad \to \mathbb{C} \times \{\text{isomorphism classes of irreducible } \mathfrak{sl}_n\text{-modules}\}\end{aligned}$$

is a bijection.

Proof. The fact that 1_n acts on V as a scalar multiplication is an example of Corollary 6.3.2. Since $\mathfrak{gl}_n = \mathbb{C}1_n \oplus \mathfrak{sl}_n$, any $\mathfrak{sl}_n$-submodule of V is also a $\mathfrak{gl}_n$-submodule of V and hence equals either $\{0\}$ or V. So V is an irreducible $\mathfrak{sl}_n$-module, as required. We thus obtain a map as in the statement, which, to any irreducible $\mathfrak{gl}_n$-module V (or rather, an isomorphism class of such modules), assigns a pair consisting of the scalar by which 1_n acts on V (clearly an isomorphism invariant of V) and the isomorphism class of irreducible $\mathfrak{sl}_n$-modules obtained by restricting the action. It is clear that, for any $a \in \mathbb{C}$ and any irreducible $\mathfrak{sl}_n$-module V, there is a unique way to extend the $\mathfrak{sl}_n$-action to $\mathfrak{gl}_n$ so that 1_n acts as multiplication by a. This gives the desired inverse of the map. □

We can also deduce from Schur's lemma an analogous result about invariant bilinear forms that generalizes Example 6.2.7.

Proposition 6.3.4. *Let V be an irreducible $\mathfrak{g}$-module.*

(1) *If $V \ncong V^*$ then there are no nonzero $\mathfrak{g}$-invariant bilinear forms on V.*
(2) *If $V \cong V^*$ then there is a unique (up to a scalar) nonzero $\mathfrak{g}$-invariant bilinear form on V; it is nondegenerate and either symmetric or skew-symmetric.*

Proof. Part (1) follows directly from part (1) of Schur's lemma, because $\mathfrak{g}$-invariant bilinear forms on V correspond to $\mathfrak{g}$-module homomorphisms $V \to V^*$. (Recall that V^* is also irreducible, by Proposition 6.1.1.) If $V \cong V^*$ then part (3) of Schur's lemma implies that the space of $\mathfrak{g}$-invariant bilinear forms is one-dimensional and so consists of all multiples of a single nonzero form, say B. Since B corresponds to an isomorphism $V \xrightarrow{\sim} V^*$, it is nondegenerate; all that remains is to prove that B is either symmetric or skew-symmetric. The transposed form B^{t} is clearly also $\mathfrak{g}$-invariant, so that $B^{\mathrm{t}} = aB$ for some $a \in \mathbb{C}$. Applying the transpose we deduce that $B = aB^{\mathrm{t}}$; thus $B = a^2 B$, which implies that either $a = 1$ (B is symmetric) or $a = -1$ (B is skew-symmetric). □

It is particularly important to consider $\mathfrak{g}$-invariant bilinear forms on $\mathfrak{g}$ itself (regarded as a $\mathfrak{g}$-module via the adjoint representation). By definition, a bilinear form $B : \mathfrak{g} \times \mathfrak{g} \to \mathbb{C}$ is invariant if it satisfies

$$B([x, y], z) + B(y, [x, z]) = 0 \quad \text{for all } x, y, z \in \mathfrak{g}. \tag{6.3.1}$$

There is a large supply of such forms.

Proposition 6.3.5. *If V is any $\mathfrak{g}$-module, the symmetric bilinear form $B_V : \mathfrak{g} \times \mathfrak{g} \to \mathbb{C} : (x, y) \mapsto \operatorname{tr}(x_V y_V)$ is $\mathfrak{g}$-invariant.*

Proof. For all $x, y, z \in \mathfrak{g}$,

$$\begin{aligned} B_V([x,y],z) + B_V(y,[x,z]) &= \operatorname{tr}([x,y]_V z_V) + \operatorname{tr}(y_V [x,z]_V) \\ &= \operatorname{tr}\Big((x_V y_V - y_V x_V) z_V + y_V (x_V z_V - z_V x_V)\Big) \\ &= \operatorname{tr}(x_V (y_V z_V)) - \operatorname{tr}((y_V z_V) x_V) = 0, \end{aligned}$$

as required. □

If $\mathfrak{g} \subseteq \mathfrak{gl}_n$, the invariant bilinear form on $\mathfrak{g}$ attached to the natural module is the *trace form* $(x, y) \mapsto \operatorname{tr}(xy)$.

Example 6.3.6. On the standard basis $\{e_{ij}\}$ of $\mathfrak{gl}_n$, the trace form is given by $\operatorname{tr}(e_{ij} e_{kl}) = \delta_{il}\delta_{jk}$: each matrix entry is 'paired' with the entry in the transpose position. Thus, in the resulting $\mathfrak{gl}_n$-module homomorphism $\mathfrak{gl}_n \to (\mathfrak{gl}_n)^*$, the basis element e_{ji} of $\mathfrak{gl}_n$ corresponds to the dual basis element e_{ij}^* of $(\mathfrak{gl}_n)^*$; in particular, this is an isomorphism and the trace form on $\mathfrak{gl}_n$ is nondegenerate. ■

Example 6.3.7. The trace form on $\mathfrak{sl}_n$ must also be nondegenerate, by Proposition 6.3.4, since $\mathfrak{sl}_n$ is an irreducible $\mathfrak{sl}_n$-module (and the trace form on $\mathfrak{sl}_n$ is clearly nonzero). Indeed, $\mathfrak{sl}_n$ is by definition the subspace of $\mathfrak{gl}_n$ perpendicular to 1_n under the trace form. ■

Example 6.3.8. The trace form on $\mathfrak{n}_n$ or any subalgebra thereof is zero. ■

In the light of Remark 4.1.9 one would expect that, even for linear Lie algebras $\mathfrak{g} \subseteq \mathfrak{gl}_n$, more intrinsic information is obtained by using the adjoint representation on $\mathfrak{g}$ rather than the natural representation on $\mathbb{C}^n$.

Definition 6.3.9. For any Lie algebra $\mathfrak{g}$, the invariant symmetric bilinear form on $\mathfrak{g}$ attached to the adjoint representation is called the *Killing form* $\kappa_{\mathfrak{g}}$. To spell out the definition,

$$\kappa_{\mathfrak{g}}(x, y) = \operatorname{tr}(\operatorname{ad}_{\mathfrak{g}}(x) \operatorname{ad}_{\mathfrak{g}}(y)) \quad \text{for all } x, y \in \mathfrak{g}.$$

Example 6.3.10. To calculate the Killing form of $\mathfrak{b}_2$, we first need the representing matrices of the adjoint representation. With respect to the basis e_{11}, e_{12}, e_{22}, these are

$$\operatorname{ad}(e_{11}) : \begin{pmatrix} 0 & 0 & 0 \\ 0 & 1 & 0 \\ 0 & 0 & 0 \end{pmatrix}, \qquad \operatorname{ad}(e_{12}) : \begin{pmatrix} 0 & 0 & 0 \\ -1 & 0 & 1 \\ 0 & 0 & 0 \end{pmatrix}, \qquad \operatorname{ad}(e_{22}) : \begin{pmatrix} 0 & 0 & 0 \\ 0 & -1 & 0 \\ 0 & 0 & 0 \end{pmatrix}.$$

Using these matrices, we calculate that

$$\begin{aligned}
\kappa_{\mathfrak{b}_2}(e_{11}, e_{11}) &= \operatorname{tr}(\operatorname{ad}(e_{11}) \operatorname{ad}(e_{11})) = 1, \\
\kappa_{\mathfrak{b}_2}(e_{11}, e_{12}) &= \operatorname{tr}(\operatorname{ad}(e_{11}) \operatorname{ad}(e_{12})) = 0, \\
\kappa_{\mathfrak{b}_2}(e_{11}, e_{22}) &= \operatorname{tr}(\operatorname{ad}(e_{11}) \operatorname{ad}(e_{22})) = -1, \\
\kappa_{\mathfrak{b}_2}(e_{12}, e_{12}) &= \operatorname{tr}(\operatorname{ad}(e_{12}) \operatorname{ad}(e_{12})) = 0, \\
\kappa_{\mathfrak{b}_2}(e_{12}, e_{22}) &= \operatorname{tr}(\operatorname{ad}(e_{12}) \operatorname{ad}(e_{22})) = 0, \\
\kappa_{\mathfrak{b}_2}(e_{22}, e_{22}) &= \operatorname{tr}(\operatorname{ad}(e_{22}) \operatorname{ad}(e_{22})) = 1.
\end{aligned}$$

Notice that this form is degenerate, since $\kappa_{\mathfrak{b}_2}(e_{12}, \cdot)$ is zero. ■

For general Lie algebras the Killing form and the extent of its degeneracy are key structural features. However, in the case of $\mathfrak{gl}_n$ and $\mathfrak{sl}_n$ the Killing form turns out not to carry significantly more information than the trace form.

Example 6.3.11. Let us calculate the Killing form of $\mathfrak{gl}_n$. We have

$$\begin{aligned}
\operatorname{ad}(e_{ij}) \operatorname{ad}(e_{kl})(e_{gh}) &= [e_{ij}, \delta_{lg} e_{kh} - \delta_{hk} e_{gl}] \\
&= \delta_{jk}\delta_{lg} e_{ih} - \delta_{hi}\delta_{lg} e_{kj} - \delta_{jg}\delta_{hk} e_{il} + \delta_{li}\delta_{hk} e_{gj}.
\end{aligned}$$

Write a_{gh} for the coefficient of e_{gh} in $\operatorname{ad}(e_{ij}) \operatorname{ad}(e_{kl})(e_{gh})$. From the above formula, we see that

$$a_{gh} = \delta_{gi}\delta_{jk}\delta_{lg} - \delta_{gk}\delta_{hj}\delta_{hi}\delta_{lg} - \delta_{gi}\delta_{hl}\delta_{jg}\delta_{hk} + \delta_{hj}\delta_{li}\delta_{hk}.$$

Hence

$$\begin{aligned}
\kappa_{\mathfrak{gl}_n}(e_{ij}, e_{kl}) &= \sum_{g,h} a_{gh} \\
&= \sum_{g,h} (\delta_{gi}\delta_{jk}\delta_{lg} - \delta_{gk}\delta_{hj}\delta_{hi}\delta_{lg} - \delta_{gi}\delta_{hl}\delta_{jg}\delta_{hk} + \delta_{hj}\delta_{li}\delta_{hk}) \\
&= 2n\delta_{il}\delta_{jk} - 2\delta_{ij}\delta_{kl}.
\end{aligned}$$

Since the right-hand side is $2n \operatorname{tr}(e_{ij}e_{kl}) - 2 \operatorname{tr}(e_{ij}) \operatorname{tr}(e_{kl})$, we have

$$\kappa_{\mathfrak{gl}_n}(x, y) = 2n \operatorname{tr}(xy) - 2 \operatorname{tr}(x) \operatorname{tr}(y) \quad \text{for all } x, y \in \mathfrak{gl}_n, \tag{6.3.2}$$

by bilinearity. This form is degenerate, because $\kappa_{\mathfrak{gl}_n}(1_n, \cdot)$ is zero. ■

Example 6.3.12. Since the trace form of $\mathfrak{sl}_n$ is nonzero, the Killing form of $\mathfrak{sl}_n$ must be a scalar multiple of it, by Proposition 6.3.4. In fact, we can deduce easily from the preceding example that the scalar is $2n$. If we use a basis of $\mathfrak{gl}_n$ that consists of 1_n followed by a basis of $\mathfrak{sl}_n$ then the matrix of any $\operatorname{ad}_{\mathfrak{gl}_n}(x)$ for $x \in \mathfrak{sl}_n$ is of the block form $\left(\begin{smallmatrix} 0 & 0 \\ 0 & * \end{smallmatrix}\right)$, where the first row and first column are all zero and the bottom-right $(n^2-1) \times (n^2-1)$ block is the matrix of $\mathrm{ad}_{\mathfrak{sl}_n}(x)$. Hence, for $x, y \in \mathfrak{sl}_n$,

$$2n\,\mathrm{tr}(xy) = \mathrm{tr}\left(\mathrm{ad}_{\mathfrak{gl}_n}(x)\,\mathrm{ad}_{\mathfrak{gl}_n}(y)\right) = \mathrm{tr}\left(\mathrm{ad}_{\mathfrak{sl}_n}(x)\,\mathrm{ad}_{\mathfrak{sl}_n}(y)\right) = \kappa_{\mathfrak{sl}_n}(x, y).$$

In particular, the Killing form of $\mathfrak{sl}_n$ is nondegenerate. ■

Remark 6.3.13. A fundamental result in the structure theory of general Lie algebras, Cartan's semisimplicity criterion, asserts that a Lie algebra $\mathfrak{g}$ has nondegenerate Killing form if and only if it has no nonzero abelian ideals. The 'only if' direction is proved in Exercise 6.5.3 below; see Fulton [4, theorem 9.9] or Humphreys [10, theorem 5.1] for the full proof.

6.4 Casimir operators

Suppose that V is a $\mathfrak{g}$-module and that we want to find submodules of V, perhaps in the hope of decomposing V completely as a direct sum of irreducible submodules. For this purpose it is useful to know a $\mathfrak{g}$-module endomorphism Ω of V because the eigenspaces of Ω are guaranteed to be $\mathfrak{g}$-submodules, by the principle used in proving part (2) of Schur's lemma (Proposition 6.3.1). Each of these eigenspaces could have further nontrivial submodules, so in general we expect to need more than just one endomorphism.

As we saw in the proof of Corollary 6.3.2, the actions of elements of the centre of $\mathfrak{g}$ are examples of such endomorphisms. The other important class of examples is given by the next result, which requires a preliminary definition.

Definition 6.4.1. Let V be a vector space, and let B be a nondegenerate bilinear form on V. For any basis $v_1, \dots, v_n$ of V, the *dual basis of V relative to B* is the basis $v_1', \dots, v_n'$ of V corresponding to the basis $v_1^*, \dots, v_n^*$ of V^* under the isomorphism $V \xrightarrow{\sim} V^* : v \mapsto B(v, \cdot)$. In other words, it is defined by the rule that $B(v_i', v_j) = \delta_{ij}$ for all i, j.

Theorem 6.4.2. *Let V be a $\mathfrak{g}$-module, and let B be a nondegenerate $\mathfrak{g}$-invariant bilinear form on $\mathfrak{g}$. Let $x_1, \dots, x_m$ be a basis of $\mathfrak{g}$, and let $x_1', \dots, x_m'$ be the dual basis of $\mathfrak{g}$ relative to B. Define a linear transformation $\Omega_{B,V}$ of V by*

$$\Omega_{B,V}(v) = \sum_{i=1}^{m} x_i x_i' v \quad \text{for all } v \in V.$$

Then $\Omega_{B,V}$ is a $\mathfrak{g}$-module endomorphism of V and is independent of the choice of basis $x_1, \dots, x_m$.

Proof. Let x be an element of $\mathfrak{g}$, and let (a_{ij}) be the matrix of $\text{ad}(x)$ with respect to the basis $x_1, \dots, x_m$, that is,

$$[x, x_i] = \sum_{j=1}^{m} a_{ji} x_j \quad \text{for } 1 \le i \le m. \tag{6.4.1}$$

The matrix of $\text{ad}(x)$ with respect to the basis $x'_1, \dots, x'_m$ is by definition the same as the matrix of the action of x on $\mathfrak{g}^*$ with respect to the basis $x^*_1, \dots, x^*_m$, which by Proposition 6.1.2 is the negative transpose of (a_{ij}). So we have

$$[x, x'_i] = -\sum_{j=1}^{m} a_{ij} x'_j \quad \text{for } 1 \le i \le m. \tag{6.4.2}$$

Hence, for all $v \in V$,

$$\begin{aligned}
x\Omega_{B,V}(v) - \Omega_{B,V}(xv) &= \sum_{i=1}^{m} x x_i x'_i v - x_i x'_i x v \\
&= \sum_{i=1}^{m} [x, x_i] x'_i v + x_i [x, x'_i] v \\
&= \sum_{i,j=1}^{m} a_{ji} x_j x'_i v - a_{ij} x_i x'_j v = 0,
\end{aligned}$$

which proves that $\Omega_{B,V}$ is a $\mathfrak{g}$-module endomorphism. To show that $\Omega_{B,V}$ does not depend on the choice of $x_1, \dots, x_m$, suppose that we use a different basis $y_1, \dots, y_m$ and its dual basis $y'_1, \dots, y'_m$ relative to B. Then, for some invertible $m \times m$ matrices (b_{ij}) and (c_{ij}), we have $y_i = \sum_{k=1}^{m} b_{ki} x_k$ and $y'_i = \sum_{k=1}^{m} c_{ki} x'_k$ for all i. Since

$$\delta_{ij} = B(y'_i, y_j) = \sum_{k,l=1}^{m} c_{li} b_{kj} B(x'_l, x_k) = \sum_{k=1}^{m} c_{ki} b_{kj},$$

the matrix (c_{ij}) is the inverse transpose of the matrix (b_{ij}). Using this, we see that, for all $v \in V$,

$$\sum_{i=1}^{m} y_i y'_i v = \sum_{i,k,l=1}^{m} b_{ki} c_{li} x_k x'_l v = \sum_{k,l=1}^{m} \delta_{kl} x_k x'_l v = \Omega_{B,V}(v),$$

as required. □

Definition 6.4.3. The endomorphism $\Omega_{B,V}$ is called the *Casimir operator* on V associated with the nondegenerate invariant bilinear form B.

Since $\Omega_{B,V}$ is built from the actions of various elements of $\mathfrak{g}$, it preserves any $\mathfrak{g}$-submodule W of V and the endomorphisms of W and V/W that it induces are clearly $\Omega_{B,W}$ and $\Omega_{B,V/W}$.

A particularly important case is that where $\mathfrak{g} = \mathfrak{gl}_n$ and B is the trace form. The basis dual to $e_{11}, e_{12}, \ldots, e_{1n}, e_{21}, \ldots, e_{nn}$ is the same basis but in the other natural order, $e_{11}, e_{21}, \ldots, e_{n1}, e_{12}, \ldots, e_{nn}$. So the Casimir operator $\Omega_V = \Omega_{\mathrm{tr},V}$ is given by

$$\Omega_V(v) = \sum_{i,j=1}^{n} e_{ij}e_{ji}v. \tag{6.4.3}$$

As always, $e_{ij}e_{ji}v$ means $e_{ij}(e_{ji}v)$ and cannot be simplified to $e_{ii}v$ (except in the case of the next example).

Example 6.4.4. Consider the natural $\mathfrak{gl}_n$-module $\mathbb{C}^n$. The Casimir operator $\Omega_{\mathbb{C}^n}$ has matrix $\sum_{i,j=1}^{n} e_{ij}e_{ji} = \sum_{i=1}^{n} ne_{ii} = n1_n$, i.e. $\Omega_{\mathbb{C}^n}$ is scalar multiplication by n. If this seems disappointing, remember that since $\mathbb{C}^n$ is an irreducible $\mathfrak{gl}_n$-module, $\Omega_{\mathbb{C}^n}$ had to be multiplication by some scalar. ■

Example 6.4.5. Consider the $\mathfrak{gl}_n$-module $\mathbb{C}^n \otimes \mathbb{C}^n$. On a standard basis element $v_i \otimes v_j$, the Casimir operator acts as follows:

$$\begin{aligned}
\Omega_{\mathbb{C}^n\otimes\mathbb{C}^n}(v_i \otimes v_j) &= \sum_{k,l=1}^{n} e_{kl}e_{lk}(v_i \otimes v_j) \\
&= \sum_{k,l=1}^{n} e_{kl}(e_{lk}v_i \otimes v_j + v_i \otimes e_{lk}v_j) \\
&= \sum_{l=1}^{n} e_{il}(v_l \otimes v_j) + \sum_{l=1}^{n} e_{jl}(v_i \otimes v_l) \\
&= \sum_{l=1}^{n}(e_{il}v_l \otimes v_j + v_l \otimes e_{il}v_j) + \sum_{l=1}^{n}(e_{jl}v_i \otimes v_l + v_i \otimes e_{jl}v_l) \\
&= n(v_i \otimes v_j) + v_j \otimes v_i + v_j \otimes v_i + n(v_i \otimes v_j).
\end{aligned}$$

So $\Omega_{\mathbb{C}^n\otimes\mathbb{C}^n} = 2n\,1_{\mathbb{C}^n\otimes\mathbb{C}^n} + 2\sigma$, where σ is the linear transformation that interchanges the two tensor factors. We see that in this case the decomposition of $\mathbb{C}^n \otimes \mathbb{C}^n$ into eigenspaces for the Casimir operator is the familiar decomposition $\mathbb{C}^n \otimes \mathbb{C}^n = \mathrm{Sym}^2(\mathbb{C}^n) \oplus \mathrm{Alt}^2(\mathbb{C}^n)$. ■

In the next chapter we will need the following property of the traces of Casimir operators.

Proposition 6.4.6. *For any $\mathfrak{gl}_n$-modules V and W,*

$$\frac{\operatorname{tr}(\Omega_{V\otimes W})}{\dim(V\otimes W)} = \frac{\operatorname{tr}(\Omega_V)}{\dim V} + \frac{\operatorname{tr}(\Omega_W)}{\dim W} + \frac{2}{n}\frac{\operatorname{tr}((1_n)_V)}{\dim V}\frac{\operatorname{tr}((1_n)_W)}{\dim W}.$$

Proof. Let $x_1, \dots, x_{n^2}$ and $x'_1, \dots, x'_{n^2}$ be bases of $\mathfrak{gl}_n$ that are dual for the trace form. Let $v_1, \dots, v_m$ and $w_1, \dots, w_p$ be bases of V and W respectively. Then, for any $1 \leq i \leq m$ and $1 \leq j \leq p$,

$$\begin{aligned}
&\Omega_{V\otimes W}(v_i \otimes w_j)\\
&= \sum_{k=1}^{n^2} x_k x'_k (v_i \otimes w_j)\\
&= \sum_{k=1}^{n^2} x_k (x'_k v_i \otimes w_j + v_i \otimes x'_k w_j)\\
&= \sum_{k=1}^{n^2} x_k x'_k v_i \otimes w_j + x'_k v_i \otimes x_k w_j + x_k v_i \otimes x'_k w_j + v_i \otimes x_k x'_k w_j\\
&= \Omega_V(v_i) \otimes w_j + v_i \otimes \Omega_W(w_j) + \sum_{k=1}^{n^2} x'_k v_i \otimes x_k w_j + x_k v_i \otimes x'_k w_j.
\end{aligned}$$

To find $\operatorname{tr}(\Omega_{V\otimes W})$ we must take the coefficient of the basis element $v_i \otimes w_j$ in this expression and sum over all i and j. The result is

$$p\operatorname{tr}(\Omega_V) + m\operatorname{tr}(\Omega_W) + \sum_{k=1}^{n^2} \operatorname{tr}((x'_k)_V)\operatorname{tr}((x_k)_W) + \operatorname{tr}((x_k)_V)\operatorname{tr}((x'_k)_W).$$

Now, suppose we choose the basis $x_1, \dots, x_{n^2}$ in such a way that $x_1, \dots, x_{n^2-1}$ is a basis of $\mathfrak{sl}_n$ and $x_{n^2} = 1_n$. Then since $\mathbb{C}1_n$ and $\mathfrak{sl}_n$ are perpendicular to each other for the trace form, we have $x'_1, \dots, x'_{n^2-1} \in \mathfrak{sl}_n$ and $x'_{n^2} = \frac{1}{n}1_n$. Moreover, since $\mathfrak{sl}_n = \mathcal{D}\mathfrak{gl}_n$ we know from Proposition 4.1.8 that $\operatorname{tr}(x_V) = \operatorname{tr}(x_W) = 0$ for all $x \in \mathfrak{sl}_n$. Hence

$$\operatorname{tr}(\Omega_{V\otimes W}) = (\dim W)\operatorname{tr}(\Omega_V) + (\dim V)\operatorname{tr}(\Omega_W) + \frac{2}{n}\operatorname{tr}((1_n)_V)\operatorname{tr}((1_n)_W),$$

from which the result follows after dividing through by $\dim(V \otimes W) = (\dim V)(\dim W)$. □

6.5 Exercises

Exercise 6.5.1. The decomposition of $V(m) \otimes V(n)$ for $m \geq n$ given in Example 6.1.10 means that, for $0 \leq k \leq n$, there is a unique (up to a scalar) highest-weight vector of weight $m+n-2k$ in $V(m) \otimes V(n)$. For instance, the highest-weight vector of weight $m+n$ is $v_0 \otimes w_0$.

(i) Express the other highest-weight vectors as linear combinations of the basis elements $v_i \otimes w_j$.
(ii) We know that the $\mathfrak{sl}_2$-submodule generated by $v_0 \otimes w_0$ must be isomorphic to $V(m+n)$. Calculate the other elements of its string basis.

Exercise 6.5.2. Show that there is no nondegenerate $\mathfrak{b}_2$-invariant bilinear form on $\mathfrak{b}_2$.

Exercise 6.5.3. Prove one direction of Cartan's semisimplicity criterion (see Remark 6.3.13), by showing that if $\mathfrak{h}$ is an abelian ideal of $\mathfrak{g}$ then $\kappa_{\mathfrak{g}}(x, y) = 0$ for all $x \in \mathfrak{h}$ and $y \in \mathfrak{g}$.

Exercise 6.5.4. Recall the construction of the semi-direct product $\mathfrak{g} \ltimes V$ from Exercise 4.5.6.

(i) Express the Killing form $\kappa_{\mathfrak{g} \ltimes V}$ in terms of the forms $\kappa_{\mathfrak{g}}$ and B_V on $\mathfrak{g}$. When is $\kappa_{\mathfrak{g} \ltimes V}$ nondegenerate?
(ii) Take $V = \mathfrak{g}^*$ (the dual of the adjoint representation). Show that the symmetric bilinear form B on $\mathfrak{g} \ltimes \mathfrak{g}^*$ defined by

$$B((x, f), (x', f')) = f(x') + f'(x) \quad \text{for } x, x' \in \mathfrak{g},\ f, f' \in \mathfrak{g}^*$$

is nondegenerate and $(\mathfrak{g} \ltimes \mathfrak{g}^*)$-invariant.

Exercise 6.5.5. We know that the Casimir operator $\Omega_{V(m)}$ on the irreducible $\mathfrak{sl}_2$-module $V(m)$, defined using the trace form on $\mathfrak{sl}_2$, must be a scalar multiplication. Find the scalar.

Exercise 6.5.6. Show that the $\mathfrak{sl}_2$-modules $\mathrm{Sym}^2(V(3))$ and $\mathrm{Sym}^3(V(2))$ are isomorphic. (*Hint*: find the weight multiplicities in each case, and use Corollary 5.2.3(3).)

Exercise 6.5.7. In this exercise we will prove that $\mathfrak{sl}_2 \times \mathfrak{sl}_2 \cong \mathfrak{so}_4$.

(i) Show that the bilinear form B_2 on $\mathbb{C}^2$ defined by $B_2((\begin{smallmatrix} a_1 \\ a_2 \end{smallmatrix}), (\begin{smallmatrix} b_1 \\ b_2 \end{smallmatrix})) = a_1 b_2 - a_2 b_1$ is $\mathfrak{sl}_2$-invariant.
(ii) Show that there is a nondegenerate symmetric bilinear form B on $\mathbb{C}^2 \otimes \mathbb{C}^2$ that satisfies

$$B(v \otimes w, v' \otimes w') = B_2(v, v') B_2(w, w') \quad \text{for all } v, v', w, w' \in \mathbb{C}^2.$$

(iii) There is an action of $\mathfrak{sl}_2 \times \mathfrak{sl}_2$ on $\mathbb{C}^2 \otimes \mathbb{C}^2$ defined by

$$(x, y)(v \otimes w) = xv \otimes w + v \otimes yw \quad \text{for all } x, y \in \mathfrak{sl}_2,\ v, w \in \mathbb{C}^2.$$

Show that the form B is $(\mathfrak{sl}_2 \times \mathfrak{sl}_2)$-invariant.

(iv) Use Proposition 6.2.6 to prove that $\mathfrak{sl}_2 \times \mathfrak{sl}_2 \cong \mathfrak{so}_4$.

Exercise 6.5.8. In this exercise we will prove that $\mathfrak{sl}_4 \cong \mathfrak{so}_6$.

(i) Show that there is a nondegenerate symmetric bilinear form B on $\mathrm{Alt}^2(\mathbb{C}^4)$ that satisfies

$$B(v\otimes w - w\otimes v, v'\otimes w' - w'\otimes v') = \det(v\ w\ v'\ w') \quad \text{for all } v, v', w, w' \in \mathbb{C}^4,$$

where $(v\ w\ v'\ w')$ denotes the 4×4 matrix with columns v, w, v', w'.

(ii) Show that B is $\mathfrak{sl}_4$-invariant.

(iii) Use Proposition 6.2.6 to prove that $\mathfrak{sl}_4 \cong \mathfrak{so}_6$.

CHAPTER 7

Integral $\mathfrak{gl}_n$-modules

In this chapter we aim to solve Problem 4.4.10 by classifying integral $\mathfrak{gl}_n$-modules. We will use the $\mathfrak{sl}_2$ theory of Chapter 5 and the tools developed in Chapter 6 (especially Casimir operators). Because of the close connection between $\mathfrak{gl}_n$-modules and $\mathfrak{sl}_n$-modules seen in Proposition 6.3.3, it is convenient to carry out the classification of $\mathfrak{sl}_n$-modules in parallel.

7.1 Integral weights

The first step is to find analogues of the basis elements e, h, f of $\mathfrak{sl}_2$, maintaining the tripartite division into strictly upper-triangular, diagonal, and strictly lower-triangular pieces. We take $n \geq 2$ throughout.

Definition 7.1.1. We will use the following notation for subalgebras of $\mathfrak{gl}_n$:

$$\begin{aligned} \mathfrak{n}^+ &= \mathbb{C}\{e_{ij} \mid 1 \leq i < j \leq n\} \quad \text{(previously called } \mathfrak{n}_n\text{)}, \\ \mathfrak{n}^- &= \mathbb{C}\{e_{ji} \mid 1 \leq i < j \leq n\}, \\ \mathfrak{d} &= \mathbb{C}\{e_{ii} \mid 1 \leq i \leq n\} \quad \text{(previously called } \mathfrak{d}_n\text{)}, \\ \mathfrak{h} &= \mathfrak{d} \cap \mathfrak{sl}_n = \left\{ \sum_{i=1}^{n} a_{ii} e_{ii} \,\middle|\, \sum_{i=1}^{n} a_{ii} = 0 \right\}. \end{aligned}$$

We name some special elements of these subalgebras:

$$e_i = e_{i,i+1} \in \mathfrak{n}^+, \ f_i = e_{i+1,i} \in \mathfrak{n}^-, \ h_i = e_{ii} - e_{i+1,i+1} \in \mathfrak{h} \text{ for } 1 \leq i \leq n-1.$$

Proposition 7.1.2. *With the notation as above:*

(1) *we have vector-space direct-sum decompositions*

$$\mathfrak{gl}_n = \mathfrak{n}^+ \oplus \mathfrak{d} \oplus \mathfrak{n}^- \quad \textit{and} \quad \mathfrak{sl}_n = \mathfrak{n}^+ \oplus \mathfrak{h} \oplus \mathfrak{n}^-;$$

(2) $\mathfrak{n}^+$ *is generated as a Lie algebra by* $e_1, \ldots, e_{n-1}$*;*

(3) $\mathfrak{n}^-$ *is generated as a Lie algebra by* $f_1, \ldots, f_{n-1}$*;*

(4) $\mathfrak{d}$ *is an abelian Lie algebra with basis* $e_{11}, \ldots, e_{nn}$ *and* $\mathfrak{h}$ *is an abelian Lie algebra with basis* $h_1, \ldots, h_{n-1}$*;*

(5) *$\mathfrak{sl}_n$ is generated as a Lie algebra by $\{e_i, h_i, f_i \mid 1 \leq i \leq n-1\}$;*

(6) *for every i, e_i, h_i, and f_i form an $\mathfrak{sl}_2$-triple, in other words*

$$[e_i, f_i] = h_i, \ [h_i, e_i] = 2e_i, \ [h_i, f_i] = -2f_i;$$

(7) *if $i \neq j$ then $[e_i, f_j] = 0$.*

Proof. Part (1) is obvious, and part (2) was proved in Example 3.1.12. Part (3) follows from part (2), because the restriction of the automorphism $x \mapsto -x^{\mathrm{t}}$ of $\mathfrak{gl}_n$ (see Exercise 2.3.7) gives an isomorphism $\mathfrak{n}^+ \xrightarrow{\sim} \mathfrak{n}^-$ that sends e_i to $-f_i$. In part (4), the statement about $\mathfrak{d}$ is obvious. To prove the statement about $\mathfrak{h}$, since $\mathfrak{h}$ is clearly $(n-1)$-dimensional it suffices to show that $h_1, \ldots, h_{n-1}$ are linearly independent, which is easy. Part (5) clearly follows from (1)–(4), and (6) and (7) hold by direct calculation. □

Definition 7.1.3. For $1 \leq i \leq n-1$, let $\mathfrak{s}_i$ be the subalgebra of $\mathfrak{sl}_n$ spanned by e_i, h_i, f_i, which is isomorphic to $\mathfrak{sl}_2$ by Proposition 7.1.2(6).

We found that the key to describing the structure of $\mathfrak{sl}_2$-modules was to consider the weight spaces, i.e. the eigenspaces of h. For $\mathfrak{gl}_n$-modules we need to consider the eigenspaces of the subalgebra $\mathfrak{d}$; according to Definition 4.3.5 these are attached to the characters of $\mathfrak{d}$, which just means the elements of the dual space $\mathfrak{d}^*$ since $\mathfrak{d}$ is abelian. We let $\varepsilon_1, \ldots, \varepsilon_n$ be the basis of $\mathfrak{d}^*$ dual to the basis $e_{11}, \ldots, e_{nn}$ of $\mathfrak{d}$; recall that this means that

$$\varepsilon_i(a_{11}e_{11} + \cdots + a_{nn}e_{nn}) = a_{ii} \quad \text{for } 1 \leq i \leq n. \tag{7.1.1}$$

Definition 7.1.4. Let V be a $\mathfrak{gl}_n$-module. For $\lambda \in \mathfrak{d}^*$, the λ-eigenspace of $\mathfrak{d}$ in V is written V_λ and called the *weight space of weight* λ. Thus

$$V_{a_1\varepsilon_1 + \cdots + a_n\varepsilon_n} = \{v \in V \mid e_{ii}v = a_i v \text{ for all } i\}.$$

A nonzero $v \in V_\lambda$ is called a *weight vector of weight* λ. Those $\lambda \in \mathfrak{d}^*$ for which $V_\lambda \neq 0$ are called the *weights* of V; $\dim V_\lambda$ is called the *multiplicity* of the weight λ in V.

Definition 7.1.5. The *weight lattice* Λ is the set of $a_1\varepsilon_1 + \cdots + a_n\varepsilon_n \in \mathfrak{d}^*$ with all $a_i \in \mathbb{Z}$. The elements of Λ are called *integral weights*.

Proposition 7.1.6. *A $\mathfrak{gl}_n$-module V is integral if and only if it has a basis $v_1, \ldots, v_m$ consisting of weight vectors whose corresponding weights $\lambda_1, \ldots, \lambda_m$ belong to Λ. If this is the case then we have a direct sum decomposition $V = \bigoplus_{\lambda \in \Lambda} V_\lambda$, and V_λ is the span of those v_i for which $\lambda_i = \lambda$. In particular, the set of weights of V equals $\{\lambda_1, \ldots, \lambda_m\}$, and the multiplicity of λ in V is simply the number of i such that $\lambda_i = \lambda$.*

Proof. Suppose that V is an integral $\mathfrak{gl}_n$-module; by definition, this means that each $(e_{ii})_V$ is diagonalizable with integer eigenvalues. Since $\mathfrak{d}$ is abelian, the linear transformations $(e_{11})_V, \dots, (e_{nn})_V$ commute with each other. By a standard linear algebra result they can be simultaneously diagonalized, so that there is a basis $v_1, \dots, v_m$ of V such that $e_{ii}v_j = a_{ij}v_j$ for some $a_{ij} \in \mathbb{Z}$. (The proof of this simultaneous diagonalization is similar to that of Proposition 4.3.8: V is a direct sum of the eigenspaces of $(e_{11})_V$ and each eigenspace W_1 is a $\mathfrak{d}$-submodule; then W_1 is a direct sum of the eigenspaces of $(e_{22})_{W_1}$ and so forth.) Set $\lambda_j = a_{1j}\varepsilon_1 + \cdots + a_{nj}\varepsilon_n \in \Lambda$. Then v_j is a weight vector of weight λ_j, as desired. Conversely, if we have a basis $v_1, \dots, v_m$ consisting of weight vectors of weights $\lambda_1, \dots, \lambda_m$ then $(e_{ii})_V$ is diagonalizable with respect to this basis, and its eigenvalues are $\lambda_1(e_{ii}), \dots, \lambda_m(e_{ii})$. If $\lambda_1, \dots, \lambda_m$ belong to Λ then by definition these eigenvalues are all integers.

Assuming that V has a basis of weight vectors as in the statement, we see that the sum of the weight spaces is the whole of V. Since weight spaces are just eigenspaces of $\mathfrak{d}$, this sum is a direct sum by Proposition 4.3.7, so $V = \bigoplus_{\lambda\in\Lambda} V_\lambda$. For any λ, let W_λ denote the span of the basis vectors v_i for which $\lambda_i = \lambda$. Clearly $W_\lambda \subseteq V_\lambda$, but we also have $V = \bigoplus_{\lambda\in\Lambda} W_\lambda$, so $W_\lambda = V_\lambda$ as claimed. The rest follows. □

To illustrate the proposition we will compute the weights and weight spaces of some basic $\mathfrak{gl}_n$-modules, verifying that they are integral. Recall from Example 4.1.19 that any character of $\mathfrak{gl}_n$ has the form $a \cdot \mathrm{tr}$ for some $a \in \mathbb{C}$.

Proposition 7.1.7. (1) *For an integer $a \in \mathbb{Z}$, the one-dimensional $\mathfrak{gl}_n$-module $\mathbb{C}_{a\cdot\mathrm{tr}}$ is integral. Its unique weight is $a(\varepsilon_1 + \cdots + \varepsilon_n)$, with multiplicity* 1.

(2) *The natural $\mathfrak{gl}_n$-module $\mathbb{C}^n$ is integral. Its weights are $\varepsilon_1, \dots, \varepsilon_n$, each with multiplicity* 1. *The weight space decomposition is*

$$\mathbb{C}^n = \mathbb{C}v_1 \oplus \cdots \oplus \mathbb{C}v_n.$$

(3) *The $\mathfrak{gl}_n$-module $(\mathbb{C}^n)^*$ is integral. Its weights are $-\varepsilon_1, \dots, -\varepsilon_n$, each with multiplicity* 1. *The weight space decomposition is*

$$(\mathbb{C}^n)^* = \mathbb{C}v_1^* \oplus \cdots \oplus \mathbb{C}v_n^*.$$

(4) *The $\mathfrak{gl}_n$-module $\mathfrak{gl}_n$ (with the adjoint representation) is integral. Its weights are* 0 *with multiplicity n and $\varepsilon_i - \varepsilon_j$, for $1 \le i \ne j \le n$, with multiplicity* 1. *The weight space decomposition is*

$$\mathfrak{gl}_n = \mathfrak{d} \oplus \bigoplus_{i\ne j} \mathbb{C}e_{ij}.$$

Proof. Part (1) is obvious. For part (2), we have $(a_{11}e_{11} + \cdots + a_{nn}e_{nn})v_i = a_{ii}v_i$, meaning that v_i is a weight vector with weight ε_i, which implies the claim. For part

(3) we have $(a_{11}e_{11} + \cdots + a_{nn}e_{nn})v_i^* = -a_{ii}v_i^*$ (a special case of Proposition 6.1.2), meaning that v_i^* is a weight vector with weight $-\varepsilon_i$, which implies the claim. For part (4), we have

$$[a_{11}e_{11} + \cdots + a_{nn}e_{nn}, e_{ij}] = (a_{ii} - a_{jj})e_{ij},$$

meaning that e_{ij} is a weight vector with weight $\varepsilon_i - \varepsilon_j$. Thus the diagonal subalgebra $\mathfrak{d}$ is the weight space of weight 0 (of dimension n), and every other weight occurs with multiplicity 1. □

It is clear from Proposition 7.1.6 that if V and W are two integral $\mathfrak{gl}_n$-modules then their direct sum $V \oplus W$ is also an integral $\mathfrak{gl}_n$-module. Moreover, we have $(V \oplus W)_\lambda = V_\lambda \oplus W_\lambda$, so that every weight of $V \oplus W$ is either a weight of V or a weight of W (and the multiplicities add). Here are some other easy properties of weights and integrality, which we will often use without comment.

Proposition 7.1.8. *Let V be an integral $\mathfrak{gl}_n$-module.*

(1) *If W is any submodule of V then W is also integral and $W_\lambda = W \cap V_\lambda$ for all $\lambda \in \Lambda$. In particular, the set of weights of W is a subset of the set of weights of V.*
(2) *The dual module V^* is also integral, and* $\dim V_\lambda^* = \dim V_{-\lambda}$ *for all λ. In particular, the weights of V^* are the negatives of the weights of V.*
(3) *If W is another integral $\mathfrak{gl}_n$-module then $V \otimes W$ is also integral and*

$$(V \otimes W)_\lambda = \bigoplus_{\mu \in \Lambda} V_\mu \otimes W_{\lambda - \mu}.$$

In particular, every weight of $V \otimes W$ is the sum of a weight of V and a weight of W.

Proof. To prove part (1), we observe that the basis $v_1, \ldots, v_m$ of V used in Proposition 7.1.6 can be chosen so that some subset of it spans W, because W is stable under all the linear transformations $(e_{ii})_V$. The result clearly follows. For part (2) we use the dual basis $v_1^*, \ldots, v_m^*$ of V^* and reason exactly as in Proposition 7.1.7(3) to show that v_i^* is a weight vector of weight equal to the negative of that of v_i. For part (3) we need to prove that if $v \in V_\mu$ and $w \in W_{\mu'}$ for $\mu, \mu' \in \Lambda$ then $v \otimes w \in (V \otimes W)_{\mu+\mu'}$. This follows from the definition of the Lie algebra action on a tensor product, because

$$\begin{aligned} d(v \otimes w) = dv \otimes w + v \otimes dw &= \mu(d)v \otimes w + v \otimes \mu'(d)w \\ &= (\mu + \mu')(d)(v \otimes w) \end{aligned}$$

for all $d \in \mathfrak{d}$. Then, if $v_1, \dots, v_m$ is a basis of weight vectors for V and $w_1, \dots, w_p$ is a basis of weight vectors for W, we have that $\{v_i \otimes w_j\}$ is a basis of weight vectors for $V \otimes W$ and the result follows. □

There is a similar theory for $\mathfrak{sl}_n$-modules, where weights and weight spaces are defined using $\mathfrak{h}$ in place of $\mathfrak{d}$. The notion of integrality is defined as follows.

Definition 7.1.9. An element $\overline{\lambda} \in \mathfrak{h}^*$ is called an *integral weight* if $\overline{\lambda}(h_i) \in \mathbb{Z}$ for all i. The set of integral weights in $\mathfrak{h}^*$ is written $\overline{\Lambda}$.

If $\varpi_1, \dots, \varpi_{n-1}$ denotes the basis of $\mathfrak{h}^*$ dual to the basis $h_1, \dots, h_{n-1}$ of $\mathfrak{h}$ then, by definition,

$$\overline{\Lambda} = \{b_1\varpi_1 + \cdots + b_{n-1}\varpi_{n-1} \mid b_i \in \mathbb{Z}\}.$$

It is often more convenient, however, to express elements of $\overline{\Lambda}$ in a different way. Namely, we have an obvious surjective linear map $\mathfrak{d}^* \to \mathfrak{h}^*$ that restricts a linear function on $\mathfrak{d}$ to the subspace $\mathfrak{h}$; the kernel of this restriction map is the one-dimensional subspace spanned by the trace function $\varepsilon_1 + \cdots + \varepsilon_n$. Thus, if we let $\overline{\varepsilon_1}, \dots, \overline{\varepsilon_n} \in \mathfrak{h}^*$ denote the restrictions of $\varepsilon_1, \dots, \varepsilon_n$ then every element of $\mathfrak{h}^*$ can be written as a linear combination $a_1\overline{\varepsilon_1} + \cdots + a_n\overline{\varepsilon_n}$. Here the n-tuple $(a_1, \dots, a_n)$ is not unique but, rather, is determined up to the addition of a scalar multiple of $(1, \dots, 1)$ (if one wanted to normalize the n-tuple definitively, the natural restriction to impose would be that $a_1 + \cdots + a_n = 0$.) Note that we have

$$a_1\overline{\varepsilon_1} + \cdots + a_n\overline{\varepsilon_n} \in \overline{\Lambda} \iff a_i - a_{i+1} \in \mathbb{Z} \quad \text{for } 1 \leq i \leq n-1. \tag{7.1.2}$$

From this it is clear that this restriction induces a surjective map $\Lambda \to \overline{\Lambda}$, i.e. every element of $\overline{\Lambda}$ can be written as a linear combination $a_1\overline{\varepsilon_1} + \cdots + a_n\overline{\varepsilon_n}$ with $a_i \in \mathbb{Z}$. (Note that in general it is not possible to achieve $a_i \in \mathbb{Z}$ and $a_1 + \cdots + a_n = 0$ simultaneously.) For example, an easy calculation shows that

$$\varpi_i = \overline{\varepsilon_1} + \cdots + \overline{\varepsilon_i} \quad \text{for } 1 \leq i \leq n-1. \tag{7.1.3}$$

If V is a $\mathfrak{gl}_n$-module and $v \in V$ is a weight vector of weight λ then, by definition, when we regard V as an $\mathfrak{sl}_n$-module, v is a weight vector of weight $\overline{\lambda}$ (where $\overline{\lambda}$ denotes λ restricted to $\mathfrak{h}$). So, if V is an integral $\mathfrak{gl}_n$-module then V regarded as an $\mathfrak{sl}_n$-module is still the direct sum of its weight spaces, with all weights integral. The remarkable fact, however, is that any $\mathfrak{sl}_n$-module at all has such a decomposition into weight spaces for integral weights.

Proposition 7.1.10. *If V is any $\mathfrak{sl}_n$-module then $V = \bigoplus_{\overline{\lambda} \in \overline{\Lambda}} V_{\overline{\lambda}}$.*

Proof. Restricting the action of $\mathfrak{sl}_n$ to $\mathfrak{s}_i$ and applying Corollary 5.2.3(1), we see that $(h_i)_V$ is diagonalizable with integer eigenvalues. Since $\mathfrak{h}$ is abelian, the linear transformations $(h_1)_V, \dots, (h_{n-1})_V$ commute with each other, and the rest of the argument is identical to that in the proof of Proposition 7.1.6. □

As a consequence, it would be tautological to speak of 'integral' $\mathfrak{sl}_n$-modules; for $\mathfrak{sl}_n$-modules, integrality is automatic.

Example 7.1.11. For $n = 2$, $\overline{\Lambda}$ is simply $\{m\overline{\varepsilon_1} \mid m \in \mathbb{Z}\}$ so we can identify it with $\mathbb{Z}$. Since $(m\varepsilon_1)(h_1) = m$, in our current terminology a weight vector of weight $m\overline{\varepsilon_1}$ in an $\mathfrak{sl}_2$-module is the same thing as a weight vector of weight m in the terminology of Definition 5.1.6. So the concepts of weights and weight spaces reduce to the corresponding concepts for $\mathfrak{sl}_2$-modules. ■

Example 7.1.12. For $n = 3$, $\overline{\Lambda}$ can be plotted in the plane. For symmetry, we put the $\overline{\varepsilon_i}$ at the vertices of an equilateral triangle:

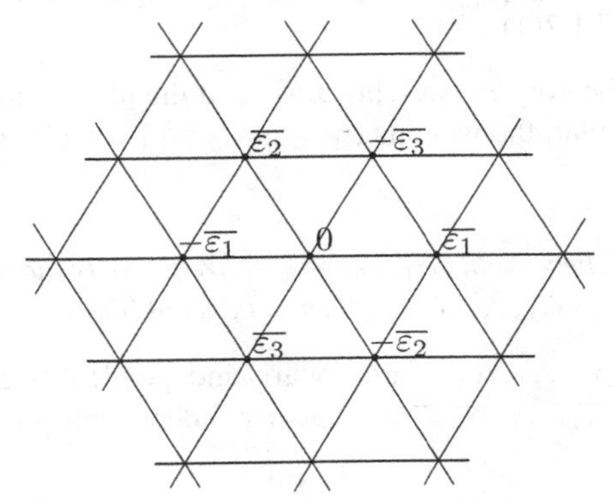

In this picture the elements of $\overline{\Lambda}$ are the intersection points of the three families of parallel lines. One can represent the weight space decomposition of an $\mathfrak{sl}_3$-module by placing small solid discs at the various weights; the size of a disc indicates the multiplicity. For instance, the $\mathfrak{sl}_3$-modules $\mathbb{C}^3$ and $(\mathbb{C}^3)^*$ would be pictured as follows:

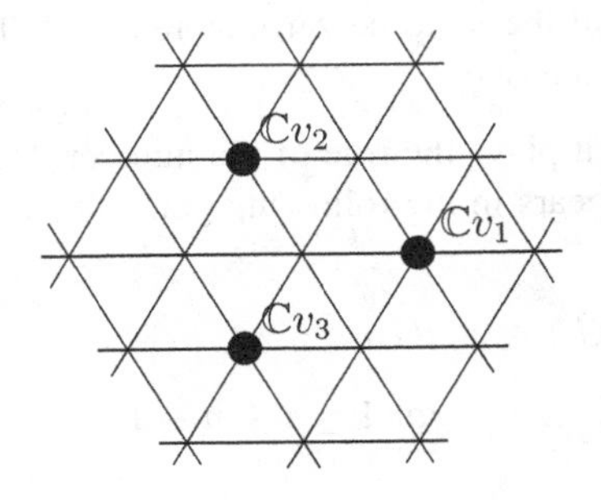

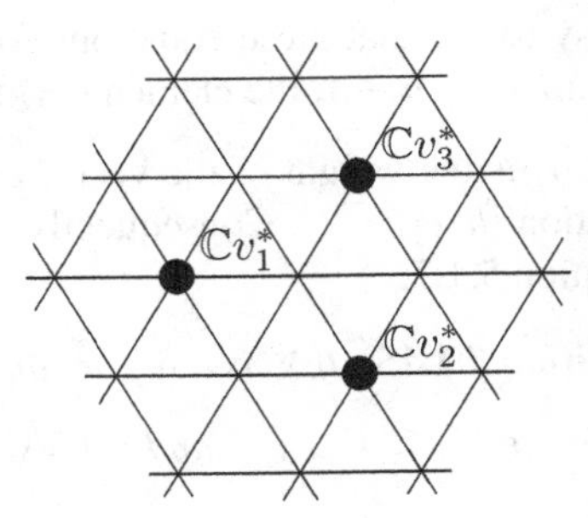

The $\mathfrak{sl}_3$-module $\mathfrak{sl}_3$ would be represented as

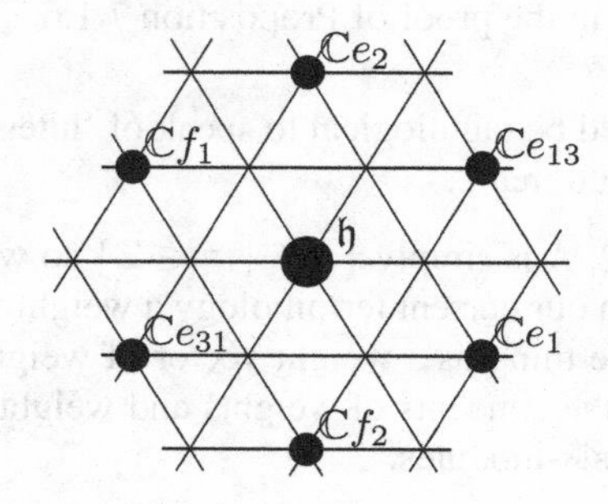

Here the larger disc at 0 indicates that $\dim \mathfrak{h} = 2$. ■

We now return to the $\mathfrak{gl}_n$-module $\mathfrak{gl}_n$, whose weight space decomposition was given in Proposition 7.1.7(4).

Definition 7.1.13. The nonzero weights $\varepsilon_i - \varepsilon_j$ of the $\mathfrak{gl}_n$-module $\mathfrak{gl}_n$ are called the *roots* of $\mathfrak{gl}_n$. In particular, the weights $\alpha_i := \varepsilon_i - \varepsilon_{i+1}$ for $1 \leq i \leq n-1$ are called the *simple roots*.

Proposition 7.1.14. *The weights $\alpha_1, \ldots, \alpha_{n-1}$ are linearly independent elements of $\mathfrak{d}^*$ and their restrictions $\overline{\alpha_1}, \ldots, \overline{\alpha_{n-1}}$ form a basis of $\mathfrak{h}^*$.*

Proof. The fact that $\alpha_1, \ldots, \alpha_{n-1}$ are linearly independent is easy. It is a stronger statement to say that $\overline{\alpha_1}, \ldots, \overline{\alpha_{n-1}}$ are linearly independent in $\mathfrak{h}^*$; to see this, note that

$$a_1\overline{\alpha_1} + \cdots + a_{n-1}\overline{\alpha_{n-1}} = a_1\overline{\varepsilon_1} + (a_2 - a_1)\overline{\varepsilon_2} + \cdots \\ + (a_{n-1} - a_{n-2})\overline{\varepsilon_{n-1}} - a_{n-1}\overline{\varepsilon_n},$$

which is zero if and only if

$$a_1 = a_2 - a_1 = \cdots = a_{n-1} - a_{n-2} = -a_{n-1}.$$

It is easy to see that these equations force all the a_i to be zero, as required. Then, since $\dim \mathfrak{h}^* = n - 1$, the elements $\overline{\alpha_i}$ form a basis. □

Now α_i is the weight of the vector e_i, so it plays the role of the number 2 in the $\mathfrak{sl}_2$ relation $[h, e] = 2e$. Consequently, it appears in the following generalization of Proposition 5.1.7.

Proposition 7.1.15. *If V is a $\mathfrak{gl}_n$-module and $\lambda \in \mathfrak{d}^*$ then*

$$e_i V_\lambda \subseteq V_{\lambda+\alpha_i} \quad \text{and} \quad f_i V_\lambda \subseteq V_{\lambda-\alpha_i} \qquad \text{for } 1 \leq i \leq n-1.$$

Proof. If $v \in V_\lambda$ and $d \in \mathfrak{d}$ then

$$de_i v = [d, e_i]v + e_i dv = (\alpha_i(d)e_i)v + e_i(\lambda(d)v) = (\lambda + \alpha_i)(d)e_i v,$$

which shows that $e_i v \in V_{\lambda+\alpha_i}$. A similar argument holds with f_i in place of e_i. □

We can deduce a beautiful symmetry of the weights of any $\mathfrak{gl}_n$-module.

Proposition 7.1.16. *If V is a $\mathfrak{gl}_n$-module, the set of weights of V, together with their multiplicities, is stable under the action of the symmetric group S_n on $\mathfrak{d}^*$, which is given by permutations of the coefficients of $\varepsilon_1, \ldots, \varepsilon_n$.*

Proof. Let $\lambda \in \mathfrak{d}^*$. From Proposition 7.1.15 we see that $W = \bigoplus_{m\in\mathbb{Z}} V_{\lambda+m\alpha_i}$ is an $\mathfrak{s}_i$-submodule of V. We can identify $\mathfrak{s}_i$ with $\mathfrak{sl}_2$ and apply to W the results about $\mathfrak{sl}_2$-modules proved in Chapter 5. Note that, since $\alpha_i(h_i) = 2$, $V_{\lambda+m\alpha_i}$ is the $(\lambda(h_i) + 2m)$-weight space of W in the terminology of Chapter 5. In particular V_λ is the $\lambda(h_i)$-weight space of W and $V_{\lambda-\lambda(h_i)\alpha_i}$ is the $(-\lambda(h_i))$-weight space of W, so Corollary 5.2.3(2) implies that

$$\dim V_\lambda = \dim V_{\lambda-\lambda(h_i)\alpha_i}. \tag{7.1.4}$$

Now, if $\lambda = a_1\varepsilon_1 + \cdots + a_n\varepsilon_n$ then

$$\begin{aligned}\lambda - \lambda(h_i)\alpha_i &= (a_1\varepsilon_1 + \cdots + a_n\varepsilon_n) - (a_i - a_{i+1})(\varepsilon_i - \varepsilon_{i+1}) \\ &= a_1\varepsilon_1 + \cdots + a_{i+1}\varepsilon_i + a_i\varepsilon_{i+1} + \cdots + a_n\varepsilon_n.\end{aligned}$$

Since the symmetric group is generated by such adjacent transpositions, we deduce that

$$\dim V_{a_1\varepsilon_1+\cdots+a_n\varepsilon_n} = \dim V_{a_{\sigma(1)}\varepsilon_1+\cdots+a_{\sigma(n)}\varepsilon_n} \quad \text{for all } \sigma \in S_n, \tag{7.1.5}$$

which is the desired result. □

We have an obvious analogue of Proposition 7.1.15 for $\mathfrak{sl}_n$: in any $\mathfrak{sl}_n$-module, e_i and f_i raise and lower weights by $\overline{\alpha_i}$. Consequently, an analogue of Proposition 7.1.16 for $\mathfrak{sl}_n$-modules, where we use the action of S_n on $\mathfrak{h}^*$ given by permuting the coefficients of $\overline{\varepsilon_1}, \ldots, \overline{\varepsilon_n}$, is also true.

Example 7.1.17. In terms of the above pictures, in the case $n = 3$ the symmetric group S_3 acts on $\overline{\Lambda}$ as the group of symmetries of the equilateral triangle with vertices $\overline{\varepsilon_1}, \overline{\varepsilon_2}, \overline{\varepsilon_3}$ (consisting of the identity, the rotations clockwise and anti-clockwise through 120°, and the three reflections in the axes of symmetry). So all pictures of $\mathfrak{sl}_3$-modules must be stable under this group of symmetries. ■

7.2 Highest-weight modules

A crucial ingredient in our classification of irreducible $\mathfrak{sl}_2$-modules was the notion of a highest-weight vector. To generalize this to $\mathfrak{gl}_n$, we need a concept of 'highest' that fits Proposition 7.1.15.

Definition 7.2.1. For $\lambda, \mu \in \mathfrak{d}^*$, we say that λ is *higher* than μ, written $\lambda \geq \mu$ or $\mu \leq \lambda$, if

$$\lambda - \mu = k_1\alpha_1 + \cdots + k_{n-1}\alpha_{n-1} \quad \text{for some } k_1, \ldots, k_{n-1} \in \mathbb{N}.$$

It is clear (using the fact that $\alpha_1, \ldots, \alpha_{n-1}$ are linearly independent) that $\leq$ is a partial order. As usual, we write $\mu < \lambda$ to mean $\mu \leq \lambda$ and $\mu \neq \lambda$.

Proposition 7.2.2. *Let* $\lambda = a_1\varepsilon_1 + \cdots + a_n\varepsilon_n, \mu = b_1\varepsilon_1 + \cdots + b_n\varepsilon_n \in \Lambda$. *Then* $\lambda \geq \mu$ *if and only if*

$$\begin{aligned} a_1 &\geq b_1, \\ a_1 + a_2 &\geq b_1 + b_2, \\ a_1 + a_2 + a_3 &\geq b_1 + b_2 + b_3, \\ &\vdots \\ a_1 + a_2 + \cdots + a_{n-1} &\geq b_1 + b_2 + \cdots + b_{n-1}, \\ a_1 + a_2 + \cdots + a_n &= b_1 + b_2 + \cdots + b_n. \end{aligned}$$

Proof. A straightforward calculation suffices. □

Definition 7.2.3. Let V be a $\mathfrak{gl}_n$-module. A weight vector $v \in V$ such that $e_i v = 0$ for all $1 \leq i \leq n-1$ is called a *highest-weight vector.*

Since the e_i generate $\mathfrak{n}^+$, it is equivalent to require that $xv = 0$ for all $x \in \mathfrak{n}^+$. In other words, a highest-weight vector is an eigenvector for $\mathfrak{d}$ that is also an invariant vector for $\mathfrak{n}^+$.

Proposition 7.2.4. *Let* V *be a nonzero integral* $\mathfrak{gl}_n$*-module. Then*

(1) *V contains a highest-weight vector;*
(2) *if $v \in V$ is a highest-weight vector of weight λ, $\lambda(h_i) \in \mathbb{N}$ for all $1 \leq i \leq n-1$.*

Proof. Since V has only finitely many weights, it must have a weight λ that is maximal for the partial order $\leq$, in the sense that no other weight μ of V satisfies $\lambda < \mu$. In particular $\lambda + \alpha_i$ is not a weight of V for any i, so Proposition 7.1.15 implies that any weight vector of weight λ must be a highest-weight vector. Since a highest-weight vector for $\mathfrak{sl}_n$ is in particular a highest-weight vector for each $\mathfrak{s}_i$, part (2) follows from $\mathfrak{sl}_2$-theory. □

Definition 7.2.5. An integral weight $\lambda \in \Lambda$ is *dominant* if $\lambda(h_i) \in \mathbb{N}$ for all i. The set of dominant integral weights is written Λ^+.

Proposition 7.2.6. *If* $\lambda = a_1\varepsilon_1 + \cdots + a_n\varepsilon_n \in \Lambda$ *then* $\lambda \in \Lambda^+$ *if and only if* $a_1 \geq a_2 \geq \cdots \geq a_n$.

Proof. This is immediate from the definition. □

Note that any S_n-orbit of integral weights contains a unique dominant weight, that in which the coefficients occur in non-increasing order.

Example 7.2.7. In $\mathbb{C}^n$ a weight vector is a highest-weight vector if and only if it is annihilated by every element of $\mathfrak{n}^+$. Clearly the unique such vector (up to a scalar) is v_1. As predicted by Proposition 7.2.4, the weight ε_1 of v_1 is dominant; in fact, it is clearly the only dominant weight among the weights $\varepsilon_1, \ldots, \varepsilon_n$ of $\mathbb{C}^n$. ■

Example 7.2.8. In $(\mathbb{C}^n)^*$ a weight vector is a highest-weight vector if and only if it vanishes on $\mathfrak{n}^+\mathbb{C}^n = \mathbb{C}\{v_1, \ldots, v_{n-1}\}$. Clearly the unique such weight vector (up to a scalar) is v_n^*. Again, its weight $-\varepsilon_n$ is the only one of the weights $-\varepsilon_1, \ldots, -\varepsilon_n$ that is dominant. ■

Example 7.2.9. In the $\mathfrak{gl}_n$-module $\mathfrak{gl}_n$, a weight vector is a highest-weight vector if and only if it commutes with every e_i. The only dominant weights of this module are 0 and $\varepsilon_1 - \varepsilon_n$. For a weight vector of weight 0, i.e. a nonzero diagonal matrix $d \in \mathfrak{d}$, the conditions $[d, e_i] = 0$ force d to be a scalar matrix. A weight vector of weight $\varepsilon_1 - \varepsilon_n$ is automatically a highest-weight vector, since there is no weight of $\mathfrak{gl}_n$ that is higher than $\varepsilon_1 - \varepsilon_n$. So the only highest-weight vectors (up to scalar) in this $\mathfrak{gl}_n$-module are 1_n and e_{1n}. Of course, there has to be a highest-weight vector in each of the submodules $\mathbb{C}1_n$ and $\mathfrak{sl}_n$, by Proposition 7.2.4. ■

We now come to the main definition of this section.

Definition 7.2.10. We say that a $\mathfrak{gl}_n$-module V is a *highest-weight module* if it is generated by a highest-weight vector.

Note that every irreducible integral $\mathfrak{gl}_n$-module must be a highest-weight module because it contains a highest-weight vector by Proposition 7.2.4, and is generated by it, by Proposition 4.3.14.

In the $\mathfrak{sl}_2$ case we exhibited a string basis for any highest-weight module and wrote down explicit formulas for the action of e, h, and f (see Proposition 5.1.10). For $\mathfrak{gl}_n$ we cannot be quite so explicit, but we have some qualitative results.

Proposition 7.2.11. *Let V be a highest-weight module generated by the highest-weight vector v of weight λ. Then*

(1) $V = \mathbb{C}\{f_{i_1} \cdots f_{i_r} v \mid r \geq 0, 1 \leq i_1, \ldots, i_r \leq n-1\}$;
(2) *if μ is a weight of V then $\mu \leq \lambda$;*
(3) $\dim V_\lambda = 1$;
(4) *if w is another highest-weight vector that generates V, $w = av$ for some $a \in \mathbb{C}^\times$.*

Proof. To prove (1) we need only show that $W = \mathbb{C}\{f_{i_1} \cdots f_{i_r} v\}$ is stable under the action of a set of generators of $\mathfrak{gl}_n$, because by assumption there is no proper $\mathfrak{gl}_n$-submodule of V containing v. It is clear that W is stable under the action of $f_1, \ldots, f_{n-1}$. By Proposition 7.1.15, $f_{i_1} \cdots f_{i_r} v$ lies in the weight space of weight $\lambda - \alpha_{i_1} - \cdots - \alpha_{i_r}$; so W is spanned by weight vectors, showing that W is stable under the action of $\mathfrak{d}$. To show that W is stable under the action of $e_1, \ldots, e_{n-1}$, we prove by induction on r the more precise statement that $e_j f_{i_1} \cdots f_{i_r} v$ is a linear combination of elements $f_{j_1} \cdots f_{j_s} v$ with $s < r$. The $r = 0$ case is the fact that $e_j v = 0$. Assuming the claim known for r, we have

$$e_j f_{i_1} f_{i_2} \cdots f_{i_{r+1}} v = [e_j, f_{i_1}] f_{i_2} \cdots f_{i_{r+1}} v + f_{i_1} e_j f_{i_2} \cdots f_{i_{r+1}} v.$$

Since $[e_j, f_{i_1}] = \delta_{j,i_1} h_j$, the first term is either 0 or a scalar multiple of $f_{i_2} \cdots f_{i_{r+1}} v$. By the induction hypothesis, $e_j f_{i_2} \cdots f_{i_{r+1}} v$ is a linear combination of elements $f_{k_1} \cdots f_{k_t} v$ with $t < r$, so the second term is a linear combination of elements $f_{i_1} f_{k_1} \cdots f_{k_t} v$ with $t < r$. So both terms have the required form and (1) is proved.

We now know that V is spanned by weight vectors with weights of the form $\lambda - \alpha_{i_1} - \cdots - \alpha_{i_r}$, so every weight of V is of this form; this proves (2). The only one of these spanning vectors that has weight λ is v itself, which proves (3). Finally, if w is a highest-weight vector of weight μ that generates V then by (2) we have both $\mu \leq \lambda$ and $\lambda \leq \mu$, so $\mu = \lambda$ and $w \in V_\lambda$; this by (3) implies that w is a scalar multiple of v. □

Thanks to Proposition 7.2.11, we can refer unambiguously to the *highest weight* of a highest-weight module V: it is the weight of any highest-weight vector that generates V, and it is higher than every other weight of V.

Example 7.2.12. Since $\mathbb{C}^n$ and $(\mathbb{C}^n)^*$ are irreducible $\mathfrak{gl}_n$-modules, they must be generated by their respective highest-weight vectors, v_1 and v_n^*. This is easy to verify directly using the rules

$$\begin{aligned} e_i v_j &= \delta_{j,i+1} v_i, \qquad f_i v_j = \delta_{j,i} v_{i+1}, \\ e_i v_j^* &= -\delta_{j,i} v_{i+1}^*, \quad f_i v_j^* = -\delta_{j,i+1} v_i^*. \end{aligned} \tag{7.2.1}$$

As predicted by Proposition 7.2.11(1), there is no need to use the e_i; successively applying various f_i to the highest-weight vector generates a basis. ■

Example 7.2.13. Since $\mathfrak{sl}_3$ is an irreducible $\mathfrak{gl}_3$-module, it follows that we must be able to generate it by successively applying f_1 and f_2 to the highest-weight vector e_{13}. Going through all the possibilities systematically, we have:

$$\begin{aligned}
\mathrm{ad}(f_1)(e_{13}) &= [f_1, e_{13}] = e_2,\\
\mathrm{ad}(f_2)(e_{13}) &= [f_2, e_{13}] = -e_1,\\
\mathrm{ad}(f_1)^2(e_{13}) &= [f_1, e_2] = 0,\\
\mathrm{ad}(f_1)\,\mathrm{ad}(f_2)(e_{13}) &= [f_1, -e_1] = h_1,\\
\mathrm{ad}(f_2)\,\mathrm{ad}(f_1)(e_{13}) &= [f_2, e_2] = -h_2,\\
\mathrm{ad}(f_2)^2(e_{13}) &= [f_2, -e_1] = 0,\\
\mathrm{ad}(f_1)^2\,\mathrm{ad}(f_2)(e_{13}) &= [f_1, h_1] = 2f_1,\\
\mathrm{ad}(f_2)\,\mathrm{ad}(f_1)\,\mathrm{ad}(f_2)(e_{13}) &= [f_2, h_1] = -f_2,\\
\mathrm{ad}(f_1)\,\mathrm{ad}(f_2)\,\mathrm{ad}(f_1)(e_{13}) &= [f_1, -h_2] = f_1,\\
\mathrm{ad}(f_2)^2\,\mathrm{ad}(f_1)(e_{13}) &= [f_2, -h_2] = -2f_2,
\end{aligned}$$

and the final basis vector e_{31} equals $\mathrm{ad}(f_2)\,\mathrm{ad}(f_1)\,\mathrm{ad}(f_2)\,\mathrm{ad}(f_1)(e_{13})$, to give just one possible way of writing it. Schematically, the action of f_1 and f_2 on the weight spaces can be represented as follows:

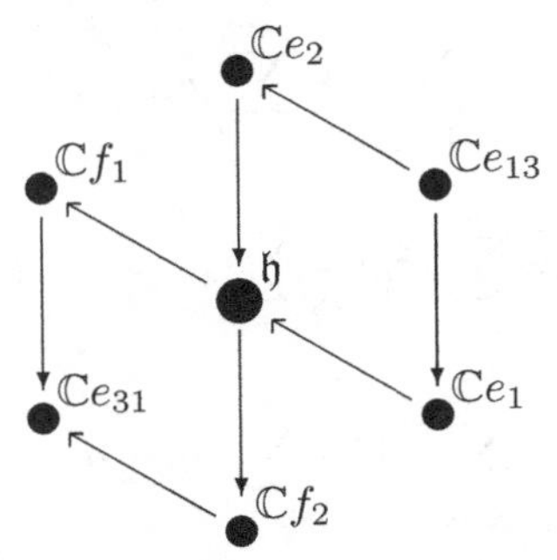

The slanting arrows represent f_1 (which lowers weights by α_1) and the vertical arrows represent f_2 (which lowers weights by α_2). ■

Every definition and result in this section has an obvious analogue for $\mathfrak{sl}_n$-modules, where we replace $\mathfrak{d}$ by $\mathfrak{h}$ and α_i by $\overline{\alpha_i}$. The set of dominant integral weights for $\mathfrak{sl}_n$ is written $\overline{\Lambda}^+$; by definition, $\overline{\Lambda}^+ = \mathbb{N}\varpi_1 + \cdots + \mathbb{N}\varpi_{n-1}$. It is clear that a highest-weight $\mathfrak{gl}_n$-module, when regarded as an $\mathfrak{sl}_n$-module, is still highest-weight.

7.3 Irreducibility of highest-weight modules

Now we come to a crucial calculation involving the Casimir operator.

Definition 7.3.1. For $\lambda = a_1\varepsilon_1 + \cdots + a_n\varepsilon_n \in \Lambda^+$, define $c_\lambda \in \mathbb{Z}$ by

$$c_\lambda = \sum_{i=1}^{n} a_i^2 + (n+1-2i)a_i.$$

Proposition 7.3.2. *If V is a highest-weight integral $\mathfrak{gl}_n$-module of highest weight λ then the Casimir operator Ω_V is scalar multiplication by c_λ.*

Proof. Let v be a highest-weight vector that generates V. By Proposition 7.2.11(1), it suffices to show that Ω_V acts as multiplication by c_λ on every vector of the form $f_{i_1}\cdots f_{i_r}v$, where $r \geq 0$ and $1 \leq i_1, \ldots, i_r \leq n-1$. But Ω_V commutes with the action of each f_i by Theorem 6.4.2. So it suffices to show that $\Omega_V(v) = c_\lambda v$. Let $\lambda = a_1\varepsilon_1 + \cdots + a_n\varepsilon_n$. Then by definition we have $e_{ii}v = a_i v$ for all i and $e_{ij}v = 0$ for all $i < j$. Thus

$$\begin{aligned}
\Omega_V(v) &= \sum_{i,j=1}^{n} e_{ij}e_{ji}v \\
&= \sum_{i=1}^{n} e_{ii}^2 v + \sum_{\substack{i,j=1\\ i<j}}^{n} e_{ij}e_{ji}v + e_{ji}e_{ij}v \\
&= \sum_{i=1}^{n} e_{ii}^2 v + \sum_{\substack{i,j=1\\ i<j}}^{n} [e_{ij}, e_{ji}]v + 2e_{ji}e_{ij}v \\
&= \sum_{i=1}^{n} e_{ii}^2 v + \sum_{\substack{i,j=1\\ i<j}}^{n} (e_{ii} - e_{jj})v + 0 \\
&= \sum_{i=1}^{n} a_i^2 v + \sum_{\substack{i,j=1\\ i<j}}^{n} (a_i - a_j)v.
\end{aligned}$$

Since each term $a_k v$ appears in the second sum $n-k$ times with a plus sign and $k-1$ times with a minus sign, its total coefficient is $n+1-2k$ and the proof is finished. □

Proposition 7.3.3. (1) *For any $\lambda \in \Lambda^+$ we have $c_\lambda \geq 0$, with equality if and only if $\lambda = 0$.*

(2) *If $\lambda, \mu \in \Lambda^+$ satisfy $\mu \leq \lambda$ then $c_\mu \leq c_\lambda$, with equality if and only if $\mu = \lambda$.*

Proof. Write $\lambda = a_1\varepsilon_1 + \cdots + a_n\varepsilon_n$ and $\mu = b_1\varepsilon_1 + \cdots + b_n\varepsilon_n$. To prove (1), recall that the dominance of λ implies that $a_1 \geq a_2 \geq \cdots \geq a_n$. Hence

$$\sum_{i=1}^{n}(n+1-2i)a_i = (n-1)(a_1 - a_n) + (n-3)(a_2 - a_{n-1}) + \cdots \geq 0,$$

which shows that

$$c_\lambda \geq \sum_{i=1}^{n} a_i^2 \geq 0,$$

with equality if and only if all the a_i are zero, as required.

For (2), we let $k_i = (a_1 + \cdots + a_i) - (b_1 + \cdots + b_i)$. By Proposition 7.2.2 the fact that $\mu \leq \lambda$ means that $k_1, \ldots, k_{n-1} \in \mathbb{N}$ and $k_n = k_0 = 0$. We have

$$\begin{aligned} c_\lambda - c_\mu &= \sum_{i=1}^{n} a_i^2 + (n+1-2i)a_i - b_i^2 - (n+1-2i)b_i \\ &= \sum_{i=1}^{n}(a_i - b_i)(a_i + b_i + n + 1 - 2i) \\ &= \sum_{i=1}^{n}(k_i - k_{i-1})(a_i + b_i + n + 1 - 2i) \\ &= \sum_{i=1}^{n-1} k_i\Big((a_i + b_i + n + 1 - 2i) - (a_{i+1} + b_{i+1} + n + 1 - 2(i+1))\Big) \\ &= \sum_{i=1}^{n-1} k_i(a_i - a_{i+1} + b_i - b_{i+1} + 2) \geq 0, \end{aligned}$$

where the last step uses $a_i \geq a_{i+1}$ and $b_i \geq b_{i+1}$. Clearly, equality holds if and only if $k_i = 0$ for all $i = 1, \ldots, n-1$, which forces $\mu = \lambda$. $\square$

These elementary facts about Casimir operators and the scalars by which they act have a major theoretical consequence:

Theorem 7.3.4. *Let V be an integral $\mathfrak{gl}_n$-module. Then V is irreducible if and only if it is a highest-weight module.*

Proof. We have seen already the reason for the 'only if' direction: if V is irreducible then it is generated by any nonzero vector that it contains, and Proposition 7.2.4 guarantees that V contains a highest-weight vector. So the substance of this result is the assertion that any highest-weight module is irreducible. We gave two proofs of the $\mathfrak{sl}_2$ case (part (2) of Theorem 5.1.15) but neither generalizes to the present situation, because we lack the required formulas for the action of the e_i on various weight spaces.

Instead, suppose that V is generated by a highest-weight vector v of weight λ. By Proposition 7.2.11(3), $V_\lambda = \mathbb{C}v$. It suffices to show that every highest-weight vector in V has weight λ, because then any nonzero $\mathfrak{gl}_n$-submodule of V must contain a nonzero multiple of v by Proposition 7.2.4, forcing it to equal V. Let $w \in V$ be a highest-weight vector of weight μ, and let W be the submodule generated by w. By Proposition 7.3.2, the Casimir operator Ω_V acts on V as scalar multiplication by c_λ and on W (where it coincides with Ω_W) as scalar multiplication by c_μ, so we must have $c_\mu = c_\lambda$. Since $\mu \leq \lambda$ by Proposition 7.2.11(2), we must have $\mu = \lambda$ by Proposition 7.3.3(2). □

Remarkably, Theorem 7.3.4 tells us that to check that an integral $\mathfrak{gl}_n$-module is irreducible, it suffices to prove that just one vector generates it, as long as that one vector is a highest-weight vector. (Compare and contrast Example 4.3.15.)

In the case of $\mathfrak{sl}_2$, Proposition 5.1.10 gave us complete formulas for the $\mathfrak{sl}_2$-action on an irreducible module in terms of the highest weight. In the present situation we do not have such formulas and it seems possible that there might be two different irreducible modules with the same highest weight. However, this can be ruled out by a wonderfully simple argument.

Proposition 7.3.5. *If V and W are two irreducible integral $\mathfrak{gl}_n$-modules of the same highest weight λ then $V \cong W$.*

Proof. Let v and w be respective highest-weight vectors for V and W. Then, in the direct sum $V \oplus W$ we have

$$d(v, w) = (dv, dw) = (\lambda(d)v, \lambda(d)w) \quad \text{for all } d \in \mathfrak{d}$$

and

$$e_i(v, w) = (e_i v, e_i w) = (0, 0) \quad \text{for } 1 \leq i \leq n-1,$$

so (v, w) is a highest-weight vector of weight λ. Let U be the submodule of $V \oplus W$ generated by (v, w). By Theorem 7.3.4, U is irreducible. The first and second projections restrict to give $\mathfrak{gl}_n$-module homomorphisms $U \to V$ and $U \to W$, which are nonzero because (v, w) is sent to v or w respectively. By Schur's lemma this implies that $U \cong V$ and $U \cong W$, so $V \cong W$. □

Thus an irreducible integral $\mathfrak{gl}_n$-module is determined up to isomorphism by its highest weight, which we know must be an element of Λ^+.

7.4 Tensor-product construction of irreducibles

The remaining question about the classification of irreducibles is whether every $\lambda \in \Lambda^+$ occurs as the highest weight of an irreducible integral $\mathfrak{gl}_n$-module. To show that it does, it suffices to construct an integral $\mathfrak{gl}_n$-module V containing a highest-weight vector v of weight λ, for then the submodule of V generated by v is irreducible by Theorem 7.3.4.

To carry out the construction, we will use tensor products.

Proposition 7.4.1. *Let $V^{(1)}, \dots, V^{(k)}$ be $\mathfrak{gl}_n$-modules. If $v^{(i)} \in V^{(i)}$ is a highest-weight vector of weight λ_i for $i = 1, \dots, k$ then $v^{(1)} \otimes \cdots \otimes v^{(k)}$ is a highest-weight vector in the $\mathfrak{gl}_n$-module $V^{(1)} \otimes \cdots \otimes V^{(k)}$ of weight $\lambda_1 + \cdots + \lambda_k$.*

Proof. By the calculation we saw in the proof of Proposition 7.1.8(3), $v^{(1)} \otimes \cdots \otimes v^{(k)}$ is a weight vector of weight $\lambda_1 + \cdots + \lambda_k$. For all $1 \leq i \leq n-1$,

$$\begin{aligned} e_i(v^{(1)} \otimes \cdots \otimes v^{(k)}) &= \sum_{j=1}^{k} v^{(1)} \otimes \cdots \otimes e_i v^{(j)} \otimes \cdots \otimes v^{(k)} \\ &= \sum_{j=1}^{k} v^{(1)} \otimes \cdots \otimes 0 \otimes \cdots \otimes v^{(k)} \\ &= \sum_{j=1}^{k} 0 \\ &= 0, \end{aligned}$$

so it is a highest-weight vector. □

Note that even if we assume that each $V^{(i)}$ is generated by the highest-weight vector $v^{(i)}$, we cannot conclude that $V^{(1)} \otimes \cdots \otimes V^{(k)}$ is a highest-weight module also; $v^{(1)} \otimes \cdots \otimes v^{(k)}$ usually generates a proper submodule, as in the following result.

Proposition 7.4.2. *The submodule of $(\mathbb{C}^n)^{\otimes k}$ generated by the highest-weight vector $v_1 \otimes \cdots \otimes v_1$ is exactly $\mathrm{Sym}^k(\mathbb{C}^n)$. So $\mathrm{Sym}^k(\mathbb{C}^n)$ is an irreducible $\mathfrak{gl}_n$-module of highest weight $k\varepsilon_1$.*

Proof. Since $v_1 \otimes \cdots \otimes v_1$ belongs to the submodule $\mathrm{Sym}^k(\mathbb{C}^n)$, the submodule that it generates is certainly contained in $\mathrm{Sym}^k(\mathbb{C}^n)$. Recall that $\mathrm{Sym}^k(\mathbb{C}^n)$ has a basis consisting of the elements

$$t_{(k_1,\dots,k_n)} = \sum_{\substack{1 \leq s_1,\dots,s_k \leq n \\ k_i \text{ of the } s_j \\ \text{equal } i}} v_{s_1} \otimes \cdots \otimes v_{s_k},$$

where $(k_1, \ldots, k_n)$ runs over all n-tuples of nonnegative integers such that $k_1 + \cdots + k_n = k$. Since each term in the sum defining $t_{(k_1,\ldots,k_n)}$ has weight

$$\varepsilon_{s_1} + \cdots + \varepsilon_{s_k} = k_1\varepsilon_1 + \cdots + k_n\varepsilon_n,$$

$t_{(k_1,\ldots,k_n)}$ is a weight vector of weight $k_1\varepsilon_1 + \cdots + k_n\varepsilon_n$. A simple calculation using (7.2.1) gives

$$f_i t_{(k_1,\ldots,k_n)} = \begin{cases} (k_{i+1}+1)\, t_{(k_1,\ldots,k_i-1,k_{i+1}+1,\ldots,k_n)} & \text{if } k_i \geq 1, \\ 0 & \text{otherwise.} \end{cases} \tag{7.4.1}$$

Hence every $t_{(k_1,\ldots,k_n)}$ can be obtained from $t_{(k,0,\ldots,0)} = v_1 \otimes \cdots \otimes v_1$ by repeatedly applying suitable f_i for various i and scaling the result. Explicitly,

$$t_{(k_1,\ldots,k_n)} = \frac{f_{n-1}^{k_n}}{k_n!} \frac{f_{n-2}^{k_{n-1}+k_n}}{(k_{n-1}+k_n)!} \cdots \frac{f_1^{k_2+\cdots+k_n}}{(k_2+\cdots+k_n)!} t_{(k,0,\ldots,0)}. \tag{7.4.2}$$

The result follows. □

Example 7.4.3. When $n = 2$, Proposition 7.4.2 implies that $\mathrm{Sym}^k(\mathbb{C}^2)$ is an irreducible $\mathfrak{sl}_2$-module of highest weight k; thus it is isomorphic to $V(k)$. The basis $t_{(k,0)}, t_{(k-1,1)}, \ldots, t_{(0,k)}$ is a string, by (7.4.2). ■

Example 7.4.4. When $n = 3$, the following table shows the results of applying f_1 and f_2 to the standard basis elements of $\mathrm{Sym}^2(\mathbb{C}^3)$.

v	$f_1 v$	$f_2 v$
$t_{(2,0,0)} = v_1 \otimes v_1$	$v_2 \otimes v_1 + v_1 \otimes v_2$	0
$t_{(1,1,0)} = v_2 \otimes v_1 + v_1 \otimes v_2$	$2(v_2 \otimes v_2)$	$v_3 \otimes v_1 + v_1 \otimes v_3$
$t_{(0,2,0)} = v_2 \otimes v_2$	0	$v_3 \otimes v_2 + v_2 \otimes v_3$
$t_{(1,0,1)} = v_3 \otimes v_1 + v_1 \otimes v_3$	$v_3 \otimes v_2 + v_2 \otimes v_3$	0
$t_{(0,1,1)} = v_3 \otimes v_2 + v_2 \otimes v_3$	0	$2(v_3 \otimes v_3)$
$t_{(0,0,2)} = v_3 \otimes v_3$	0	0

We also know that there is a submodule of $\mathbb{C}^3 \otimes \mathbb{C}^3$ complementary to $\mathrm{Sym}^2(\mathbb{C}^3)$, namely $\mathrm{Alt}^2(\mathbb{C}^3)$. The following table shows the action of f_1 and f_2 on the standard basis of this submodule.

v	$f_1 v$	$f_2 v$
$u_{1,2} = v_1 \otimes v_2 - v_2 \otimes v_1$	0	$v_1 \otimes v_3 - v_3 \otimes v_1$
$u_{1,3} = v_1 \otimes v_3 - v_3 \otimes v_1$	$v_2 \otimes v_3 - v_3 \otimes v_2$	0
$u_{2,3} = v_2 \otimes v_3 - v_3 \otimes v_2$	0	0

Thus the $\mathfrak{gl}_3$-module $\mathrm{Alt}^2(\mathbb{C}^3)$ is generated by the highest-weight vector $u_{1,2}$ of weight $\varepsilon_1 + \varepsilon_2$ and is therefore irreducible. So the decomposition $\mathbb{C}^3 \otimes \mathbb{C}^3 = \mathrm{Sym}^2(\mathbb{C}^3) \oplus \mathrm{Alt}^2(\mathbb{C}^3)$ is a decomposition of $\mathbb{C}^3 \otimes \mathbb{C}^3$ as a direct sum of irreducible $\mathfrak{gl}_3$-modules. Considering the weights for the restriction to $\mathfrak{sl}_3$, this decomposition can be depicted as follows:

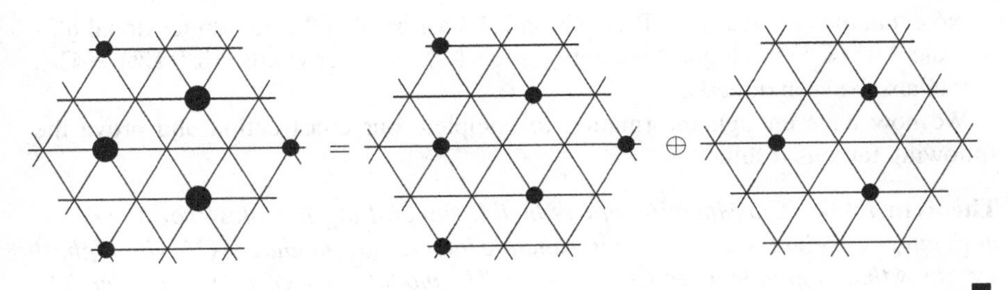

■

Generalizing the above calculations in $\mathrm{Alt}^2(\mathbb{C}^3)$ gives:

Proposition 7.4.5. *For any* $k \leq n$, $\mathrm{Alt}^k(\mathbb{C}^n)$ *is generated by*

$$u_{1,2,\ldots,k} = \sum_{\sigma \in S_k} \varepsilon(\sigma)\, v_{\sigma(1)} \otimes v_{\sigma(2)} \otimes \cdots \otimes v_{\sigma(k)},$$

which is a highest-weight vector of weight $\varepsilon_1 + \cdots + \varepsilon_k$. *So* $\mathrm{Alt}^k(\mathbb{C}^n)$ *is an irreducible* $\mathfrak{gl}_n$*-module of highest weight* $\varepsilon_1 + \cdots + \varepsilon_k$.

Proof. Recall that $\mathrm{Alt}^k(\mathbb{C}^n)$ has a basis consisting of the elements

$$u_{i_1,\ldots,i_k} = \sum_{\sigma \in S_k} \varepsilon(\sigma)\, v_{i_{\sigma(1)}} \otimes v_{i_{\sigma(2)}} \otimes \cdots \otimes v_{i_{\sigma(k)}} \qquad \text{for } 1 \leq i_1 < \cdots < i_k \leq n.$$

It is clear that $u_{i_1,\ldots,i_k}$ is a weight vector of weight $\varepsilon_{i_1} + \cdots + \varepsilon_{i_k}$. To show that $u_{1,\ldots,k}$ is a highest-weight vector, we observe (bearing in mind (7.2.1)) that $e_j u_{1,\ldots,k} = 0$ for $j \geq k$. For $1 \leq j \leq k-1$, e_j takes each term in the sum defining $u_{1,\ldots,k}$ to a pure tensor with factors $v_1, \ldots, v_j, v_j, v_{j+2}, \ldots, v_k$ (in some particular order). In the whole sum $e_j u_{1,\ldots,k}$, each such pure tensor occurs twice with opposite signs, so its total coefficient is zero (indeed, its coefficient is zero in every element of $\mathrm{Alt}^k(\mathbb{C}^n)$). Thus $e_j u_{1,\ldots,k} = 0$ as required. By similar reasoning, we see that,

$$f_j u_{i_1,\ldots,i_k} = \begin{cases} u_{i_1,\ldots,i_{s-1},j+1,i_{s+1},\ldots,i_k} & \text{if } j = i_s,\ j+1 \neq i_{s+1}, \\ 0 & \text{if } j = i_s,\ j+1 = i_{s+1}, \\ 0 & \text{if } j \neq i_s \text{ for all } s. \end{cases} \tag{7.4.3}$$

It is clear from this that every basis element $u_{i_1,\dots,i_k}$ can be obtained by applying suitable f_j to $u_{1,\dots,k}$. Explicitly,

$$u_{i_1,\dots,i_k} = (f_{i_1-1}f_{i_1-2}\cdots f_1)\cdots(f_{i_k-1}f_{i_k-2}\cdots f_k)\,u_{1,\dots,k}. \tag{7.4.4}$$

The result follows. □

Note that the $k = n$ case of Proposition 7.4.5 asserts that the one-dimensional $\mathfrak{gl}_n$-module $\mathrm{Alt}^n(\mathbb{C}^n)$ has highest weight $\varepsilon_1 + \cdots + \varepsilon_n$. In other words $\mathrm{Alt}^n(\mathbb{C}^n) \cong \mathbb{C}_{\mathrm{tr}}$, as can also be seen directly.

We now have enough information to complete our construction and prove the following famous result.

Theorem 7.4.6. *[Classification of irreducible integral $\mathfrak{gl}_n$-modules] For any dominant integral weight λ, there is an irreducible integral $\mathfrak{gl}_n$-module $V(\lambda)$ with highest weight λ that is unique up to isomorphism. The modules $V(\lambda)$, as λ runs over Λ^+, form a complete list of irreducible integral $\mathfrak{gl}_n$-modules up to isomorphism, with no isomorphism classes repeated.*

Proof. Let $\lambda = a_1\varepsilon_1 + \cdots + a_n\varepsilon_n \in \Lambda^+$. We have

$$\begin{aligned}\lambda = {} & (a_1 - a_2)\varepsilon_1 + (a_2 - a_3)(\varepsilon_1 + \varepsilon_2) + \cdots \\ & + (a_{n-1} - a_n)(\varepsilon_1 + \cdots + \varepsilon_{n-1}) + a_n(\varepsilon_1 + \cdots + \varepsilon_n).\end{aligned}$$

By the definition of Λ^+ we have $a_i - a_{i+1} \in \mathbb{N}$ for $1 \le i \le n-1$, and $a_n \in \mathbb{Z}$. So we have an integral $\mathfrak{gl}_n$-module

$$\begin{aligned}& \underbrace{\mathbb{C}^n \otimes \cdots \otimes \mathbb{C}^n}_{a_1 - a_2 \text{ factors}} \otimes \underbrace{\mathrm{Alt}^2(\mathbb{C}^n) \otimes \cdots \otimes \mathrm{Alt}^2(\mathbb{C}^n)}_{a_2 - a_3 \text{ factors}} \otimes \cdots \\ & \qquad \otimes \underbrace{\mathrm{Alt}^{n-1}(\mathbb{C}^n) \otimes \cdots \otimes \mathrm{Alt}^{n-1}(\mathbb{C}^n)}_{a_{n-1} - a_n \text{ factors}} \otimes \mathbb{C}_{a_n \cdot \mathrm{tr}}\end{aligned}$$

that contains a highest-weight vector of weight λ. The submodule generated by this highest-weight vector is irreducible by Theorem 7.3.4; this is our desired $V(\lambda)$. Its uniqueness was proved in Proposition 7.3.5. Moreover, we know from Theorem 7.3.4 and Proposition 7.2.11 that every irreducible integral $\mathfrak{gl}_n$-module has a unique highest weight λ and is therefore isomorphic to a unique $V(\lambda)$. □

We already know how to describe some irreducible modules $V(\lambda)$:

$$\begin{aligned} V(0) &\cong \mathbb{C}, \\ V(a(\varepsilon_1 + \cdots + \varepsilon_n)) &\cong \mathbb{C}_{a\cdot\mathrm{tr}}, \\ V(\varepsilon_1) &\cong \mathbb{C}^n, \\ V(-\varepsilon_n) &\cong (\mathbb{C}^n)^*, \end{aligned}$$

$$V(\varepsilon_1 - \varepsilon_n) \cong \mathfrak{sl}_n,$$
$$V(k\varepsilon_1) \cong \mathrm{Sym}^k(\mathbb{C}^n),$$
$$V(\varepsilon_1 + \cdots + \varepsilon_k) \cong \mathrm{Alt}^k(\mathbb{C}^n).$$

We can derive others by using the following rules.

Proposition 7.4.7. *Let* $\lambda = a_1\varepsilon_1 + \cdots + a_n\varepsilon_n \in \Lambda^+$.

(1) $V(\lambda)^* \cong V(\lambda^*)$, *where* $\lambda^* = -a_n\varepsilon_1 - \cdots - a_1\varepsilon_n$.
(2) *For* $a \in \mathbb{Z}$, $V(\lambda) \otimes \mathbb{C}_{a\cdot\mathrm{tr}} \cong V((a_1 + a)\varepsilon_1 + \cdots + (a_n + a)\varepsilon_n)$.

Proof. For (1) recall that $V(\lambda)^*$ is irreducible by Proposition 6.1.1. In view of Propositions 7.1.8(2) and 7.1.16, the map

$$\Lambda \to \Lambda : \quad b_1\varepsilon_1 + \cdots + b_n\varepsilon_n \mapsto -b_n\varepsilon_1 - \cdots - b_1\varepsilon_n$$

induces a bijection from the weights of $V(\lambda)$ to the weights of $V(\lambda)^*$. This map respects the partial order $\leq$, as may be seen from Proposition 7.2.2. So it must take the highest weight of $V(\lambda)$ to the highest weight of $V(\lambda)^*$, as claimed. For (2), $V(\lambda) \otimes \mathbb{C}_{a.\,\mathrm{tr}}$ is irreducible as seen in Example 6.1.9. By Proposition 7.1.8(3), the weights of $V(\lambda) \otimes \mathbb{C}_{a\cdot\mathrm{tr}}$ are obtained from the weights of $V(\lambda)$ by adding $a(\varepsilon_1 + \cdots + \varepsilon_n)$ to the latter. This operation clearly respects the partial order $\leq$, so the highest weight of $V(\lambda) \otimes \mathbb{C}_{a\cdot\mathrm{tr}}$ is $\lambda + a(\varepsilon_1 + \cdots + \varepsilon_n)$, which proves the claim. □

Notice that 1_n acts on $V(a_1\varepsilon_1 + \cdots + a_n\varepsilon_n)$ as multiplication by the scalar $a_1 + \cdots + a_n$ (since that is how it acts on the highest-weight vector). Tensoring with the one-dimensional $\mathfrak{gl}_n$-module $\mathbb{C}_{a\cdot\mathrm{tr}}$ changes the scalar by which 1_n acts but leaves the $\mathfrak{sl}_n$-action unaffected and therefore gives nothing substantially new.

Other irreducible modules $V(\lambda)$ can be constructed as submodules of tensor products, thanks to Proposition 7.4.1. Note that for this construction we can use any expression of λ as a sum of highest weights of known $\mathfrak{gl}_n$-modules: the particular recipe given in the proof of Theorem 7.4.6 is just one possibility.

Example 7.4.8. Writing $2\varepsilon_1 - \varepsilon_3$ as $(2\varepsilon_1) + (-\varepsilon_3)$, we see that the $\mathfrak{gl}_3$-module $V(2\varepsilon_1 - \varepsilon_3)$ can be constructed as the submodule of $\mathrm{Sym}^2(\mathbb{C}^3) \otimes (\mathbb{C}^3)^*$ generated by the highest-weight vector $v = t_{(2,0,0)} \otimes v_3^*$. Using (7.2.1) and (7.4.1), we find:

$$f_1 v = t_{(1,1,0)} \otimes v_3^*,$$
$$f_2 v = -t_{(2,0,0)} \otimes v_2^*,$$
$$f_2 f_1 v = t_{(1,0,1)} \otimes v_3^* - t_{(1,1,0)} \otimes v_2^*,$$
$$f_1 f_2 v = -t_{(1,1,0)} \otimes v_2^* + t_{(2,0,0)} \otimes v_1^*.$$

We could carry on such calculations to find a spanning set, and hence a basis, for the whole submodule. However, it is perhaps more enlightening to restrict ourselves to finding the weights of $V(2\varepsilon_1 - \varepsilon_3)$ and their multiplicities, which we can do more elegantly. If $a\varepsilon_1 + b\varepsilon_2 + c\varepsilon_3$ is a weight of $V(2\varepsilon_1 - \varepsilon_3)$ then we have

$$a \leq 2, \quad a + b \leq 2, \quad a + b + c = 1,$$

by Propositions 7.2.2 and 7.2.11(2). Moreover, the same conditions must hold for any permutation of (a, b, c), by Proposition 7.1.16. It is easy to see that the set of weights this allows is as follows:

$$(S_3\text{-orbit of } 2\varepsilon_1 - \varepsilon_3) \cup (S_3\text{-orbit of } \varepsilon_1 + \varepsilon_2 - \varepsilon_3) \cup (S_3\text{-orbit of } \varepsilon_1).$$

We also know that the multiplicity of the weight $2\varepsilon_1 - \varepsilon_3$ is 1, since the corresponding weight space is the span of the highest-weight vector v. By Proposition 7.1.16 again, the multiplicity of every other weight in the S_3-orbit of $2\varepsilon_1 - \varepsilon_3$ must also be 1. Since f_i lowers the weight by α_i, the only vector of the form $f_{i_1} f_{i_2} \cdots f_{i_r} v$ that has weight $\varepsilon_1 + \varepsilon_2 - \varepsilon_3$ is $f_1 v$. Since we calculated that $f_1 v$ is nonzero, the multiplicity of the weight $\varepsilon_1 + \varepsilon_2 - \varepsilon_3$, and hence of every weight in its S_3-orbit, is 1. Finally, the only vectors of the form $f_{i_1} f_{i_2} \cdots f_{i_r} v$ that have weight ε_1 are $f_1 f_2 v$ and $f_2 f_1 v$. Since our calculations show that $f_1 f_2 v$ and $f_2 f_1 v$ are linearly independent, the multiplicity of the weight ε_1, and hence of every weight in its S_3-orbit, is 2. In particular, $\dim V(2\varepsilon_1 - \varepsilon_3) = 6 \times 1 + 3 \times 1 + 3 \times 2 = 15$. The weight space decomposition of $V(2\varepsilon_1 - \varepsilon_3)$ (or rather, the $\mathfrak{sl}_3$-module obtained from it by restricting the action) can be depicted as follows:

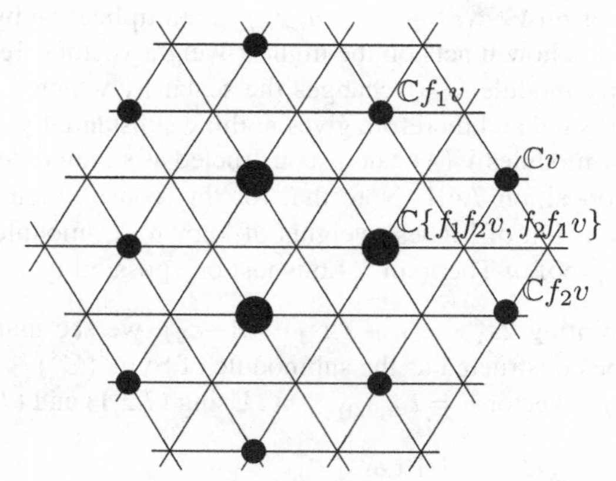

■

In view of Proposition 6.3.3 we can derive from Theorem 7.4.6 the classification of irreducible $\mathfrak{sl}_n$-modules.

Theorem 7.4.9. *For any $\overline{\lambda} \in \overline{\Lambda}^+$, there is an irreducible $\mathfrak{sl}_n$-module $V(\overline{\lambda})$ with highest weight $\overline{\lambda}$, unique up to isomorphism. The modules $V(\overline{\lambda})$, as $\overline{\lambda}$ runs over*

$\overline{\Lambda}^+$, *form a complete list of irreducible $\mathfrak{sl}_n$-modules up to isomorphism, with no isomorphism classes repeated.*

Proof. Given $\overline{\lambda} \in \overline{\Lambda}^+$, we can clearly find $\lambda \in \Lambda^+$ whose restriction to $\mathfrak{h}$ is $\overline{\lambda}$. Then the $\mathfrak{gl}_n$-module $V(\lambda)$, regarded as an $\mathfrak{sl}_n$-module, is a highest-weight module with highest weight $\overline{\lambda}$ and is irreducible by Proposition 6.3.3; this is our desired $V(\overline{\lambda})$. The proof that every irreducible $\mathfrak{sl}_n$-module V is a highest-weight module with uniquely determined highest weight $\overline{\lambda} \in \overline{\Lambda}^+$ is the same as for $\mathfrak{gl}_n$; we then have an isomorphism $V \cong V(\overline{\lambda})$ by the same argument as was used in proving Proposition 7.3.5. □

Note that, when $n = 2$, Theorem 7.4.9 agrees with Theorem 5.1.15: the irreducible $\mathfrak{sl}_2$-module $V(m\overline{\varepsilon_1})$ is what we previously called $V(m)$.

7.5 Complete reducibility

To complete the generalization of the results in Chapter 5, we want to prove that every $\mathfrak{sl}_n$-module and integral $\mathfrak{gl}_n$-module is completely reducible. We will start with what seems like a very special case.

Proposition 7.5.1. *Let V be an $\mathfrak{sl}_n$-module and W an irreducible submodule of codimension* 1 *in V. Then there is a submodule W' of V such that $V = W \oplus W'$.*

Proof. By Theorem 7.4.9, we know that $W \cong V(\overline{\lambda})$ for some $\overline{\lambda} \in \overline{\Lambda}^+$. Write $\overline{\lambda}$ as $a_1\overline{\varepsilon_1}+\cdots+a_n\overline{\varepsilon_n}$ where $a_1, \ldots, a_n$ are normalized by the requirement that $a_1+\cdots+a_n = 0$. Then the a_i are rational numbers such that $a_i - a_{i+1} \in \mathbb{N}$ for $1 \leq i \leq n-1$, but we do not necessarily have $a_i \in \mathbb{Z}$. Extend the $\mathfrak{sl}_n$-action on V to $\mathfrak{gl}_n$ by declaring that 1_n acts as zero. Then W is a $\mathfrak{gl}_n$-submodule of V, which is generated over $\mathfrak{sl}_n$ by a highest-weight vector v of weight $\overline{\lambda}$. Since $1_n v = 0$, v is also a highest-weight vector for the $\mathfrak{gl}_n$-action, of weight $\lambda = a_1\varepsilon_1 + \cdots + a_n\varepsilon_n$. By Proposition 7.3.2 (which does not actually require that $a_i \in \mathbb{Z}$), Ω_W is scalar multiplication by c_λ.

Since V/W is one-dimensional and $\mathcal{D}\mathfrak{sl}_n = \mathfrak{sl}_n$, the action of $\mathfrak{sl}_n$ on V/W is trivial; hence the action of the whole of $\mathfrak{gl}_n$ on V/W is trivial. In particular, $\Omega_{V/W} = 0$. So the eigenvalues of Ω_V are c_λ and 0. If $c_\lambda \neq 0$ then W is the c_λ-eigenspace of Ω_V and the 0-eigenspace of Ω_V is a complementary subspace W', which is an $\mathfrak{sl}_n$-submodule since Ω_V commutes with the action of $\mathfrak{sl}_n$. If $c_\lambda = 0$ then $\lambda = 0$ by Proposition 7.3.3(1) (which again does not require $a_i \in \mathbb{Z}$). So $W \cong V(0) \cong \mathbb{C}$, meaning that V is two-dimensional. Choose a basis v_1, v_2 of V such that $W = \mathbb{C}v_1$; then the representing matrices for the action of $\mathfrak{gl}_n$ all have the form $\left(\begin{smallmatrix} 0 & * \\ 0 & 0 \end{smallmatrix}\right)$. Since all such matrices commute with each other, $\mathfrak{sl}_n = \mathcal{D}\mathfrak{gl}_n$ must act as zero so $W' = \mathbb{C}v_2$ is a complementary $\mathfrak{sl}_n$-submodule. □

The next result removes the restriction that W is irreducible.

Proposition 7.5.2. *Let V be an $\mathfrak{sl}_n$-module and W a submodule of codimension* 1 *in V. Then there is a submodule W' such that $V = W \oplus W'$.*

Proof. The proof is by induction on $\dim V$: the case $\dim V = 1$ is obvious, so assume that $\dim V \geq 2$. If W is irreducible, invoke Proposition 7.5.1; otherwise, W has some nontrivial submodule U. By the induction hypothesis applied to $W/U \subset V/U$, there is a submodule X' of V/U such that $V/U = W/U \oplus X'$. The preimage of X' in V is a submodule $\widetilde{W}$ of V, containing U, such that $\dim \widetilde{W}/U = \dim X' = 1$, $V = W + \widetilde{W}$, and $W \cap \widetilde{W} = U$. Now, by the induction hypothesis applied to $U \subset \widetilde{W}$ there is a submodule W' of $\widetilde{W}$ such that $\widetilde{W} = U \oplus W'$. It follows that $V = W \oplus W'$, as required. □

It is now surprisingly easy to deduce complete reducibility for general $\mathfrak{sl}_n$-modules, using a trick due to Brauer.

Theorem 7.5.3. *Every $\mathfrak{sl}_n$-module V is completely reducible.*

Proof. Let W be any nontrivial submodule of V. Recall from Proposition 6.2.1 that we have an $\mathfrak{sl}_n$-action on the space $\mathrm{Hom}(V, W)$ of linear maps from V to W. Let $\mathcal{V}$ be the subspace of $\mathrm{Hom}(V, W)$ consisting of those linear maps $\varphi : V \to W$ for which the restriction $\varphi|_W$ is a scalar multiplication, and let $\mathcal{W}$ be the subspace of $\mathcal{V}$ consisting of those φ for which $\varphi|_W = 0$. We have $x\mathcal{V} \subseteq \mathcal{W}$ for all $x \in \mathfrak{sl}_n$, since if $\varphi|_W = a1_W$ and $w \in W$ then

$$(x\varphi)(w) = x\varphi(w) - \varphi(xw) = axw - axw = 0.$$

So $\mathcal{V}$ is an $\mathfrak{sl}_n$-submodule of $\mathrm{Hom}(V, W)$ and $\mathcal{W}$ is an $\mathfrak{sl}_n$-submodule of $\mathcal{V}$. Moreover it is clear that $\dim \mathcal{V}/\mathcal{W} = 1$, so by Proposition 7.5.2 there is a one-dimensional submodule $\mathbb{C}\varphi$ of $\mathcal{V}$ that is complementary to $\mathcal{W}$. Since $\mathcal{D}\mathfrak{sl}_n = \mathfrak{sl}_n$ the action of $\mathfrak{sl}_n$ on $\mathbb{C}\varphi$ is trivial, which means that $\varphi : V \to W$ is an $\mathfrak{sl}_n$-module homomorphism. The fact that $\varphi \notin \mathcal{W}$ means that $\varphi|_W$ is a nonzero scalar multiplication, so $\mathrm{im}(\varphi) = W$ and $\ker(\varphi) \cap W = \{0\}$. Thus $\ker(\varphi)$ is a submodule complementary to W. □

Corollary 7.5.4. *Every integral $\mathfrak{gl}_n$-module V is completely reducible.*

Proof. Since V has a basis of weight vectors, V is the direct sum of the eigenspaces of $(1_n)_V$, which are $\mathfrak{gl}_n$-submodules because $1_n \in Z(\mathfrak{gl}_n)$. So we may assume that there is only one such eigenspace, i.e. $(1_n)_V$ is a scalar multiplication. In this case, however, the $\mathfrak{gl}_n$-submodules of V are the same as the $\mathfrak{sl}_n$-submodules of V, so the result follows from Theorem 7.5.3. □

Corollary 7.5.5. *Let V be an integral $\mathfrak{gl}_n$-module.*

(1) *The module V can be written as a direct sum $V_1 \oplus \cdots \oplus V_s$ where each V_i is an irreducible submodule, isomorphic to $V(\lambda_i)$ for a unique $\lambda_i \in \Lambda^+$.*

(2) *For any $\lambda \in \Lambda^+$ the number of indices i such that $\lambda_i = \lambda$ equals $\dim(V_\lambda \cap V^{\mathfrak{n}^+})$, and the sum $V_{[\lambda]}$ of all the corresponding V_i equals the submodule of V generated by $V_\lambda \cap V^{\mathfrak{n}^+}$. In particular, both are independent of the chosen direct sum decomposition of V.*

Proof. The fact that V is completely reducible implies, by Theorem 4.4.5, that V can be written as a direct sum $V_1 \oplus \cdots \oplus V_s$ of irreducible submodules. By the classification of irreducibles each V_i is isomorphic to a unique $V(\lambda_i)$, so (1) is proved. For (2), let $\lambda \in \Lambda^+$. We have

$$V_\lambda \cap V^{\mathfrak{n}^+} = ((V_1)_\lambda \cap V_1^{\mathfrak{n}^+}) \oplus \cdots \oplus ((V_s)_\lambda \cap V_s^{\mathfrak{n}^+}).$$

By definition, however, $(V_i)_\lambda \cap V_i^{\mathfrak{n}^+}$ is nonzero if and only if there is a highest-weight vector of weight λ in V_i, which occurs if and only if $V_i \cong V(\lambda)$, i.e. $\lambda_i = \lambda$. Moreover in this case $\dim((V_i)_\lambda \cap V_i^{\mathfrak{n}^+}) = 1$, so $\dim(V_\lambda \cap V^{\mathfrak{n}^+})$ equals the number of i such that $\lambda_i = \lambda$, as claimed. The claim about $V_{[\lambda]}$ follows from the fact that, whenever $\lambda_i = \lambda$, V_i is generated by $(V_i)_\lambda \cap V_i^{\mathfrak{n}^+}$. □

The analogue of Corollary 7.5.5 for $\mathfrak{sl}_n$-modules clearly also holds.

Part (2) of Corollary 7.5.5 means that we have a well-defined concept of the multiplicity of an irreducible $V(\lambda)$ as a direct summand of a given integral $\mathfrak{gl}_n$-module V. Moreover, that multiplicity equals $\dim(V_\lambda \cap V^{\mathfrak{n}^+})$; in particular, $V(\lambda)$ occurs as a summand of V if and only if V contains a highest-weight vector of weight λ. Note that the decomposition of V into irreducible submodules is in general not unique: when the multiplicity of $V(\lambda)$ is greater than 1, the different ways to write $V_\lambda \cap V^{\mathfrak{n}^+}$ as a direct sum of one-dimensional subspaces generate different ways to write $V_{[\lambda]}$ as a direct sum of copies of $V(\lambda)$.

Remark 7.5.6. As mentioned in Chapter 1, a $\mathfrak{gl}_n$-module is integral if and only if the $\mathfrak{gl}_n$-action can be 'integrated' to give a representation of the Lie group GL_n; the proof of the 'if' direction was given there. By Corollary 7.5.5, to prove the 'only if' direction one can assume that the module is $V(\lambda)$ for some $\lambda \in \Lambda^+$. In the proof of Theorem 7.4.6 we constructed $V(\lambda)$ as the $\mathfrak{gl}_n$-submodule of a certain tensor product generated by a particular highest-weight vector. For each factor of the tensor product, it is easy to see that the $\mathfrak{gl}_n$-action can be integrated to GL_n. From this one can deduce that the same holds for the tensor product and for its submodule $V(\lambda)$.

In a theoretical sense, Corollary 7.5.5 completes our solution of Problem 4.4.10. However, as Exercise 6.5.1 showed, finding highest-weight vectors of weights that are not maximal for the partial order $\leq$ is not a trivial matter and identifying the submodules that they generate can be even more difficult. In some cases, the Casimir operator provides helpful information.

Proposition 7.5.7. *Let V be an integral $\mathfrak{gl}_n$-module. Then Ω_V is diagonalizable and its eigenvalues are precisely the nonnegative integers c_λ for $\lambda \in \Lambda^+$ such that V has a highest-weight vector of weight λ. For any $c \in \mathbb{N}$, the c-eigenspace of Ω_V is the sum of those $V_{[\lambda]}$ such that $c_\lambda = c$.*

Proof. This follows from Proposition 7.3.2 and Corollary 7.5.5. □

The problem that one encounters when using Proposition 7.5.7 is that it is possible to have different $\lambda, \mu \in \Lambda^+$ such that $c_\lambda = c_\mu$ (see Exercise 7.6.1). So an eigenspace of Ω_V could contain irreducible submodules of more than one isomorphism class. However, there is an important case for which we know that one eigenspace of Ω_V is irreducible, namely the case of a tensor product of irreducible modules.

Proposition 7.5.8. *Let $V = V(\lambda_1) \otimes \cdots \otimes V(\lambda_k)$ for some $\lambda_1, \ldots \lambda_k \in \Lambda^+$. Let $v^{(i)}$ denote a highest-weight vector in $V(\lambda_i)$.*

(1) *Every weight μ of V satisfies $\mu \leq \lambda_1 + \cdots + \lambda_k$, and the weight space $V_{\lambda_1+\cdots+\lambda_k}$ is one-dimensional and spanned by $v^{(1)} \otimes \cdots \otimes v^{(k)}$.*
(2) *The largest eigenvalue of Ω_V is $c_{\lambda_1+\cdots+\lambda_k}$.*
(3) *The $c_{\lambda_1+\cdots+\lambda_k}$-eigenspace of Ω_V equals the submodule generated by $v^{(1)} \otimes \cdots \otimes v^{(k)}$, which is isomorphic to $V(\lambda_1 + \cdots + \lambda_k)$.*
(4) *We have*

$$\frac{\operatorname{tr}(\Omega_V)}{\dim V} = c_{\lambda_1} + \cdots + c_{\lambda_k} + \frac{2}{n} \sum_{1 \leq i < j \leq k} \lambda_i(1_n)\lambda_j(1_n).$$

Proof. We can form a basis of V by taking pure tensors where the factors range over fixed weight bases in $V(\lambda_1), \ldots, V(\lambda_k)$. These pure tensors are weight vectors of weight $\mu_1 + \cdots + \mu_k$, where each μ_i is a weight of $V(\lambda_i)$. Since $\mu_i \leq \lambda_i$, we have $\mu_1 + \cdots + \mu_k \leq \lambda_1 + \cdots + \lambda_k$, with equality only in the case that $\mu_i = \lambda_i$. This proves (1). Hence the eigenvalues of Ω_V are of the form c_μ, where $\mu \in \Lambda^+$ satisfies $\mu \leq \lambda_1 + \cdots + \lambda_k$. By Proposition 7.3.3(2) the largest of these eigenvalues is $c_{\lambda_1+\cdots+\lambda_k}$ and all the others are strictly smaller. This proves (2) and (3). We prove (4) by induction on k. The $k = 1$ case says that

$$\frac{\operatorname{tr}(\Omega_{V(\lambda_1)})}{\dim V(\lambda_1)} = c_{\lambda_1},$$

which follows from Proposition 7.3.2. Assuming the claim to be true for $k-1$, we apply Proposition 6.4.6 to find that

$$\begin{aligned}\frac{\operatorname{tr}(\Omega_V)}{\dim V} &= \frac{\operatorname{tr}(\Omega_{V(\lambda_1)\otimes\cdots\otimes V(\lambda_{k-1})})}{\dim(V(\lambda_1)\otimes\cdots\otimes V(\lambda_{k-1}))} + \frac{\operatorname{tr}(\Omega_{V(\lambda_k)})}{\dim V(\lambda_k)} \\ &\quad + \frac{2}{n}\frac{\operatorname{tr}((1_n)_{V(\lambda_1)\otimes\cdots\otimes V(\lambda_{k-1})})}{\dim(V(\lambda_1)\otimes\cdots\otimes V(\lambda_{k-1}))}\frac{\operatorname{tr}((1_n)_{V(\lambda_k)})}{\dim V(\lambda_k)} \\ &= c_{\lambda_1} + \cdots + c_{\lambda_{k-1}} + \frac{2}{n}\sum_{1\le i<j\le k-1}\lambda_i(1_n)\lambda_j(1_n) + c_{\lambda_k} \\ &\quad + \frac{2}{n}(\lambda_1(1_n) + \cdots + \lambda_{k-1}(1_n))\lambda_k(1_n),\end{aligned}$$

since 1_n acts on $V(\lambda_i)$ as scalar multiplication by $\lambda_i(1_n)$ and hence on $V(\lambda_1)\otimes\cdots\otimes V(\lambda_{k-1})$ as scalar multiplication by $\lambda_1(1_n)+\cdots+\lambda_{k-1}(1_n)$. This completes the induction. □

To illustrate Proposition 7.5.8, we describe another class of irreducible $\mathfrak{gl}_n$-modules. Note that $V(\varepsilon_1-\varepsilon_n)\cong\mathfrak{sl}_n$ may be defined as the kernel of the linear map $\mathbb{C}^n\otimes(\mathbb{C}^n)^*\to\mathbb{C}$ that sends $v_i\otimes v_j^*$ to δ_{ij}, since this corresponds to the trace map under the canonical isomorphism $\mathbb{C}^n\otimes(\mathbb{C}^n)^*\cong\mathfrak{gl}_n$. We can generalize this using the basis of $\operatorname{Sym}^k(\mathbb{C}^n)$ defined in Proposition 7.4.2.

Proposition 7.5.9. *For any $k\ge 2$, define a linear map*

$$\varphi : \operatorname{Sym}^k(\mathbb{C}^n)\otimes(\mathbb{C}^n)^*\to\operatorname{Sym}^{k-1}(\mathbb{C}^n)$$

by the rule

$$\varphi(t_{(k_1,\dots,k_n)}\otimes v_j^*) = \begin{cases} t_{(k_1,\dots,k_j-1,\dots,k_n)} & \text{if } k_j\ge 1,\\ 0 & \text{otherwise.}\end{cases}$$

Then φ is a $\mathfrak{gl}_n$-module homomorphism and $\ker(\varphi)\cong V(k\varepsilon_1-\varepsilon_n)$.

Proof. Since the linear map $\mathbb{C}^n\otimes(\mathbb{C}^n)^*\to\mathbb{C}$ that sends $v_i\otimes v_j^*$ to δ_{ij} is a $\mathfrak{gl}_n$-module homomorphism, so is the linear map $\psi : (\mathbb{C}^n)^{\otimes k}\otimes(\mathbb{C}^n)^*\to(\mathbb{C}^n)^{\otimes k-1}$ defined by

$$\psi(v_{i_1}\otimes\cdots\otimes v_{i_k}\otimes v_j^*) = \frac{1}{k}\sum_{a=1}^{k}\delta_{i_a j}\,v_{i_1}\otimes\cdots\otimes\widehat{v_{i_a}}\otimes\cdots\otimes v_{i_k},$$

where the caret indicates that the corresponding factor is not present. It is easy to check that ψ restricted to $V = \mathrm{Sym}^k(\mathbb{C}^n) \otimes (\mathbb{C}^n)^*$ coincides with the map φ, which is therefore also a $\mathfrak{gl}_n$-module homomorphism.

Let $W = \ker(\varphi)$. Now, the image of φ is the whole of $\mathrm{Sym}^{k-1}(\mathbb{C}^n)$ since it contains a basis of that space. So $V/W \cong \mathrm{Sym}^{k-1}(\mathbb{C}^n) \cong V((k-1)\varepsilon_1)$, which means that the Casimir operator $\Omega_{V/W}$ is scalar multiplication by $c_{(k-1)\varepsilon_1}$. By Proposition 7.5.8(2), every eigenvalue of Ω_V is bounded above by $c_{k\varepsilon_1-\varepsilon_n}$. Hence

$$\begin{aligned}
\mathrm{tr}(\Omega_V) &\leq c_{k\varepsilon_1-\varepsilon_n} \dim W + c_{(k-1)\varepsilon_1} \dim V/W \\
&= c_{k\varepsilon_1-\varepsilon_n} \dim V - (c_{k\varepsilon_1-\varepsilon_n} - c_{(k-1)\varepsilon_1}) \dim V/W \\
&= \Big(k^2 + 1 + (n-1)(k+1)\Big) \dim V - 2(n+k-1)\binom{n+k-2}{k-1} \\
&= (c_{k\varepsilon_1} + c_{-\varepsilon_n}) \dim V - 2k\binom{n+k-1}{k} \\
&= \left(c_{k\varepsilon_1} + c_{-\varepsilon_n} - \frac{2k}{n}\right) \dim V.
\end{aligned}$$

However, Proposition 7.5.8(4) tells us that $\mathrm{tr}(\Omega_V)$ equals this last quantity. We conclude that the $c_{k\varepsilon_1-\varepsilon_n}$-eigenspace of Ω_V equals W, so $W \cong V(k\varepsilon_1 - \varepsilon_n)$ by Proposition 7.5.8(3). □

Example 7.5.10. In Example 7.4.8 we found the weight multiplicities of the $\mathfrak{gl}_3$-module $V(2\varepsilon_1 - \varepsilon_3)$, constructed as the submodule of $\mathrm{Sym}^2(\mathbb{C}^3) \otimes (\mathbb{C}^3)^*$ generated by the highest-weight vector $t_{(2,0,0)} \otimes v_3^*$. Proposition 7.5.9 tells us that this submodule is the kernel of the surjective $\mathfrak{gl}_3$-module homomorphism φ : $\mathrm{Sym}^2(\mathbb{C}^3) \otimes (\mathbb{C}^3)^* \to \mathbb{C}^3$. It follows that we have a direct sum decomposition $\mathrm{Sym}^2(\mathbb{C}^3) \otimes (\mathbb{C}^3)^* = V(2\varepsilon_1 - \varepsilon_3) \oplus W'$, where $W' \cong \mathbb{C}^3$. This gives us another way to find the weight multiplicities of $V(2\varepsilon_1 - \varepsilon_3)$. ■

7.6 Exercises

Exercise 7.6.1. Show that if V is an irreducible integral $\mathfrak{gl}_n$-module then Ω_V and Ω_{V^*} are multiplications by the same scalar.

Exercise 7.6.2. Deduce from Proposition 7.4.7 that, for any $1 \leq k \leq n-1$, we have an isomorphism $\mathrm{Alt}^{n-k}(\mathbb{C}^n) \cong \mathrm{Alt}^k(\mathbb{C}^n)^* \otimes \mathbb{C}_{\mathrm{tr}}$ of $\mathfrak{gl}_n$-modules.

Exercise 7.6.3. Imitating Example 7.4.8, find the weight multiplicities of the $\mathfrak{gl}_3$-module $V(3\varepsilon_1 - \varepsilon_3)$. Check your answer using Proposition 7.5.9.

Exercise 7.6.4. As shown in Example 7.5.10, the $\mathfrak{gl}_3$-module $\mathrm{Sym}^2(\mathbb{C}^3) \otimes (\mathbb{C}^3)^*$ has a submodule isomorphic to $\mathbb{C}^3$. Find a basis for this submodule.

Exercise 7.6.5. Show that if the integral $\mathfrak{gl}_n$-modules V and W have the same weight multiplicities, i.e. $\dim V_\lambda = \dim W_\lambda$ for all $\lambda \in \Lambda$, then $V \cong W$. (*Hint*: use induction on the dimension.)

Exercise 7.6.6. Use Proposition 7.5.8 to determine which $\lambda, \mu \in \Lambda^+$ have the property that $V(\lambda) \otimes V(\mu)$ is an irreducible $\mathfrak{gl}_n$-module.

Exercise 7.6.7. Imitate Proposition 7.5.9 to give a description of the irreducible $\mathfrak{gl}_n$-module $V(\varepsilon_1 + \cdots + \varepsilon_k - \varepsilon_n)$, for any $k \le n-1$.

Exercise 7.6.8. Write the $\mathfrak{gl}_n$-module $\mathbb{C}^n \otimes \mathbb{C}^n \otimes \mathbb{C}^n$ as a direct sum of irreducible $\mathfrak{gl}_n$-submodules. (*Hint*: use $\mathbb{C}^n \otimes \mathbb{C}^n = \mathrm{Sym}^2(\mathbb{C}^n) \oplus \mathrm{Alt}^2(\mathbb{C}^n)$ and the known submodules $\mathrm{Sym}^3(\mathbb{C}^n)$ and $\mathrm{Alt}^3(\mathbb{C}^n)$ of $\mathbb{C}^n \otimes \mathbb{C}^n \otimes \mathbb{C}^n$.)

Exercise 7.6.9. If V is a $\mathfrak{gl}_3$-module, define a linear transformation by

$$\Psi_V(v) = \sum_{i_1,i_2,i_3 \in \{1,2,3\}} e_{i_1 i_2} e_{i_2 i_3} e_{i_3 i_1} v \quad \text{for all } v \in V.$$

(i) Show that Ψ_V is a $\mathfrak{gl}_3$-module endomorphism of V.

(ii) If V is irreducible with highest weight $a_1\varepsilon_1 + a_2\varepsilon_2 + a_3\varepsilon_3 \in \Lambda^+$ then Ψ_V must be scalar multiplication. Find the scalar.

CHAPTER 8

Guide to further reading

Having arrived at this point, the reader should be ready to progress to more advanced textbooks on Lie algebras, Lie groups, and their representations. For the theory of Lie groups, which we touched on briefly in Chapter 1, good starting points are the books by Carter, Segal and Macdonald [3], Hall [8], and Rossmann [15]; those by Bourbaki [1] and Knapp [13] provide more detailed accounts. The aim of this chapter is to indicate a few high points of the theory of Lie algebras that are now in prospect and to provide some specific references that the reader can consult in order to reach them.

8.1 Classification of simple Lie algebras

The only simple Lie algebras that we have studied in this book are the special linear Lie algebras $\mathfrak{sl}_n$. As mentioned in Remark 3.3.11, there is a complete classification of simple Lie algebras due to Cartan and Killing. This is a main topic in many introductory books on Lie algebras, for example those by Erdmann and Wildon [4] and Humphreys [10]. A crucial fact that emerges in this classification is that every simple Lie algebra is generated by $\mathfrak{sl}_2$-triples, as we saw for $\mathfrak{sl}_n$ in Proposition 7.1.2. One can think of simple Lie algebras as 'molecules' made from 'atoms' that are all identical (the only 'element' being $\mathfrak{sl}_2$), and the problem is then to determine the different possibilities for the 'bonds' between the atoms.

Example 8.1.1. To illustrate this, we define a 10-dimensional Lie algebra

$$\mathfrak{g} = \left\{ \sum_{i,j} a_{ij} e_{ij} \in \mathfrak{gl}_5 \mid a_{6-j,6-i} = -a_{ij} \text{ for all } 1 \leq i, j \leq 5 \right\}.$$

That is, $\mathfrak{g}$ consists of the 5×5 matrices that are skew-symmetric not about the usual diagonal but about the diagonal running from bottom left to top right. More intrinsically, we have $\mathfrak{g} = \{x \in \mathfrak{gl}_5 \,|\, xB = 0\}$ where B is the symmetric nondegenerate bilinear form on $\mathbb{C}^5$ defined by $B(v_i, v_j) = \delta_{i,6-j}$, and the action of $\mathfrak{gl}_5$ on $\mathrm{Bil}(\mathbb{C}^5)$ is as in (6.2.2). This second description makes it easy to see that $\mathfrak{g}$ is a subalgebra of $\mathfrak{gl}_5$ and (using Proposition 6.2.6) that $\mathfrak{g} \cong \mathfrak{so}_5$.

We obtain a direct sum decomposition $\mathfrak{g} = \mathfrak{n}^+ \oplus \mathfrak{h} \oplus \mathfrak{n}^-$, analogous to that of $\mathfrak{sl}_n$, by defining $\mathfrak{h}$ to be the abelian subalgebra consisting of diagonal elements of $\mathfrak{g}$, $\mathfrak{n}^+$ the subalgebra of upper-triangular elements, and $\mathfrak{n}^-$ the subalgebra of lower-triangular elements. Note that this would not work for $\mathfrak{so}_5$, which is why we are considering $\mathfrak{g}$ instead.

Imitating Definition 7.1.1 as closely as possible, we find two $\mathfrak{sl}_2$-triples in $\mathfrak{g}$:

$$e_1 = e_{12} - e_{45},\ h_1 = e_{11} - e_{22} + e_{44} - e_{55},\ \ f_1 = e_{21} - e_{54}$$

and

$$e_2 = e_{23} - e_{34},\ h_2 = 2e_{22} - 2e_{44},\ \ f_2 = 2e_{32} - 2e_{43}.$$

(The coefficients of 2 in h_2 and f_2 are forced by the $\mathfrak{sl}_2$-relations.) It is easy to check that $\mathfrak{n}^+$ is generated by e_1 and e_2, that $\mathfrak{n}^-$ is generated by f_1 and f_2, and that $\mathfrak{h}$ has basis h_1, h_2; thus $\mathfrak{g}$ as a whole is generated by these two $\mathfrak{sl}_2$-triples. Hence we can define a weight space decomposition for any $\mathfrak{g}$-module by considering common eigenvectors for h_1, h_2, as in the case of $\mathfrak{sl}_3$. Once again, e_1 and e_2 are weight vectors in the $\mathfrak{g}$-module $\mathfrak{g}$ and the corresponding weights $\alpha_1, \alpha_2 \in \mathfrak{h}^*$, called the *simple roots* of $\mathfrak{g}$, are linearly independent. It follows that the 0-weight space in $\mathfrak{g}$ is just $\mathfrak{h}$, as in the case of $\mathfrak{sl}_3$.

However, these $\mathfrak{sl}_2$-triples in $\mathfrak{g}$ are 'bonded' differently from those we found in $\mathfrak{sl}_3$. Specifically, we have $\alpha_1(h_2) = -2$ (as opposed to -1 in $\mathfrak{sl}_3$), which implies that $[e_2, e_1]$ and $[e_2, [e_2, e_1]]$ are both nonzero (whereas the latter vanishes in $\mathfrak{sl}_3$). Consequently, the *roots* of $\mathfrak{g}$, i.e. the nonzero weights of $\mathfrak{g}$ considered as a $\mathfrak{g}$-module, are

$$\pm\alpha_1,\quad \pm\alpha_2,\quad \pm(\alpha_1 + \alpha_2),\quad \pm(\alpha_1 + 2\alpha_2).$$

Each root has multiplicity 1 as a weight of $\mathfrak{g}$, in accordance with the fact that $\dim \mathfrak{g} = \dim \mathfrak{h} + 8 = 10$. Using the weight space decomposition of $\mathfrak{g}$, it is not hard to prove that an ideal of $\mathfrak{g}$ contained in $\mathfrak{h}$ must be the zero ideal and that an ideal of $\mathfrak{g}$ not contained in $\mathfrak{h}$ must be the whole of $\mathfrak{g}$ (the argument is similar to that in the proof of Theorem 3.2.11, which used implicitly the weight space decomposition of $\mathfrak{gl}_n$). Hence $\mathfrak{g}$ is simple. ■

The first step in analysing an arbitrary simple Lie algebra $\mathfrak{g}$ is to show that it has a subalgebra $\mathfrak{h}$ with the properties we have seen in the cases of $\mathfrak{sl}_n$ and Example 8.1.1:

- $\mathfrak{h}$ is abelian;
- $\mathfrak{g}$ is the direct sum of the weight spaces (i.e. the eigenspaces for $\mathfrak{h}$);
- the 0-weight space is $\mathfrak{h}$;
- every root (i.e. nonzero weight) has multiplicity 1.

In Humphreys' book [10] this is proved in section 8, using several results from earlier chapters. Since it adds no extra difficulty, Humphreys allows $\mathfrak{g}$ to be merely *semisimple*, meaning that it has no nonzero abelian ideals. Other authors such as Carter [2] take a different approach, first introducing the general notion of a *Cartan subalgebra* and then showing that when $\mathfrak{g}$ is semisimple any Cartan subalgebra $\mathfrak{h}$ has the above properties.

The focus then shifts to the set of roots of $\mathfrak{g}$. It turns out that one can always find a basis $\alpha_1, \dots, \alpha_\ell$ of $\mathfrak{h}^*$ consisting of *simple roots* of which every other root is an integer linear combination, either with all nonnegative coefficients or all nonpositive coefficients; this dichotomy gives a notion of *positive* and *negative roots* and thus a direct sum decomposition $\mathfrak{g} = \mathfrak{n}^+ \oplus \mathfrak{h} \oplus \mathfrak{n}^-$.

For each $i \in \{1, \dots, \ell\}$, there is an essentially unique $\mathfrak{sl}_2$-triple e_i, h_i, f_i in $\mathfrak{g}$ where e_i has weight α_i, h_i belongs to $\mathfrak{h}$, and f_i has weight $-\alpha_i$. The Lie algebra $\mathfrak{g}$ is generated by these $\mathfrak{sl}_2$-triples, and the isomorphism class of $\mathfrak{g}$ is determined purely by the *Cartan matrix* $(\alpha_i(h_j))_{i,j=1}^{\ell}$, which specifies how the $\mathfrak{sl}_2$-triples are 'bonded'. See [10, section 14]. Further work is then needed to show that the Cartan matrix does not depend on the choice of $\mathfrak{h}$ and $\alpha_1, \dots, \alpha_\ell$, up to a reordering of the rows and columns; see [2, section 6.1] or [10, section 16]. In particular, the number of generating $\mathfrak{sl}_2$-triples, namely $\ell = \dim \mathfrak{h}$, is an invariant of $\mathfrak{g}$ called the *rank*.

What remains is to determine which matrices arise as the Cartan matrix of a simple Lie algebra. There are some obvious constraints, obtained by regarding $\mathfrak{g}$ as a module for $\mathbb{C}\{e_i, h_i, f_i\} \cong \mathfrak{sl}_2$ and applying Corollary 5.2.3; these are encapsulated in the axioms of a *root system*, as given in [10, section 9]. The classification of root systems is purely combinatorial and independent of Lie algebras (indeed, it arises also in the theory of finite reflection groups). The result is a list of possible Cartan matrices, bearing the following names. In each case the subscript indicates the size of the matrix:

$$A_\ell\ (\ell \geq 1), \quad B_\ell\ (\ell \geq 2), \quad C_\ell\ (\ell \geq 3), \quad D_\ell\ (\ell \geq 4),$$
$$E_\ell\ (\ell = 6, 7, 8), \quad F_4, \quad G_2.$$

See [10, section 11] for these matrices and their graphical representation by *Dynkin diagrams*.

The Cartan–Killing classification states that each matrix in the list does indeed arise as the Cartan matrix of a simple Lie algebra, so that the isomorphism classes of simple Lie algebras of rank ℓ are in bijection with the $\ell \times \ell$ matrices in the list. To illustrate, consider the 2×2 matrices,

$$A_2 = \begin{pmatrix} 2 & -1 \\ -1 & 2 \end{pmatrix}, \qquad B_2 = \begin{pmatrix} 2 & -2 \\ -1 & 2 \end{pmatrix}, \qquad G_2 = \begin{pmatrix} 2 & -1 \\ -3 & 2 \end{pmatrix}.$$

The first two are the Cartan matrices of $\mathfrak{sl}_3$ and $\mathfrak{so}_5$ respectively, as seen in Chapter 7 and Example 8.1.1. If $\mathfrak{g}$ is a simple Lie algebra with Cartan matrix G_2 then $\mathfrak{g}$ must be generated by six elements $e_1, h_1, f_1, e_2, h_2, f_2$ that satisfy the following Lie bracket relations:

$$\begin{gathered}
[e_1, f_1] = h_1, \quad [h_1, e_1] = 2e_1, \quad [h_1, f_1] = -2f_1, \\
[e_2, f_2] = h_2, \quad [h_2, e_2] = 2e_2, \quad [h_2, f_2] = -2f_2, \\
[h_1, h_2] = 0, \quad [h_1, e_2] = -3e_2, \quad [h_1, f_2] = 3f_2, \\
[f_1, e_2] = 0, \quad [e_1, f_2] = 0, \\
[e_1, [e_1, [e_1, [e_1, e_2]]]] = [f_1, [f_1, [f_1, [f_1, f_2]]]] = 0, \\
[h_2, e_1] = -e_1, \quad [h_2, f_1] = f_1, \\
[e_2, [e_2, e_1]] = [f_2, [f_2, f_1]] = 0.
\end{gathered} \tag{8.1.1}$$

For instance, $[h_1, f_2] = 3f_2$ holds because $\alpha_2(h_1) = -3$ and $[e_1, f_2] = 0$ holds because $\alpha_1 - \alpha_2$ is not a root (having both positive and negative coefficients); together these imply that f_2 is a highest-weight vector of weight 3 for the $\mathfrak{sl}_2$-triple e_1, h_1, f_1, from which it follows that $[f_1, [f_1, [f_1, [f_1, f_2]]]] = 0$.

One can construct such a Lie algebra explicitly as $\mathrm{Der}(\mathfrak{C})$, where $\mathfrak{C}$ is the non-associative Cayley–Dickson algebra [10, section 19], or by defining a suitable Lie bracket on $\mathfrak{sl}_3 \oplus \mathbb{C}^3 \oplus (\mathbb{C}^3)^*$; see the book by Fulton and Harris [6], section 22.2. Alternatively, one can define $\mathfrak{g}$ abstractly by means of the above generators and relations and then show that $\mathfrak{g}$ is finite-dimensional and simple as expected. The latter approach works in the general case and is the basis of Serre's proof of the classification, given in [10, section 18]. From this viewpoint, simple finite-dimensional Lie algebras are a special case of the class of *Kac–Moody algebras* studied in the books by Carter [2] and Kac [12].

8.2 Representations of simple Lie algebras

The representation theory of a general simple Lie algebra $\mathfrak{g}$ is much like that of $\mathfrak{sl}_n$ (or $\mathfrak{gl}_n$, in the integral case). Indeed, once the generating $\mathfrak{sl}_2$-triples $\{e_i, h_i, f_i \,|\, 1 \leq i \leq \ell\}$ have been specified, as in the previous section, many definitions and results in Chapter 7 carry over verbatim. In particular, we have the theorem of the highest weight, generalizing Theorem 7.4.9: the isomorphism classes of irreducible $\mathfrak{g}$-modules are in bijection with the dominant integral weights, via the map that associates with an irreducible $\mathfrak{g}$-module its highest weight. For enlightening pictures and heuristic discussion, consult Fulton and Harris [6, part III]; complete proofs are contained in the books by Carter [2], Hall [8], and Humphreys [10].

The reader will notice some differences between the proofs in these references and those in Chapter 7. The reason is that restricting attention to $\mathfrak{sl}_n$ and $\mathfrak{gl}_n$ simplified our arguments, in two main respects.

The first is that we were able to use the Casimir operator $\Omega_V = \Omega_{\mathrm{tr},V}$, which, when V is a highest-weight module, acts by an easily computable scalar (see Proposition 7.3.2). There is an adequate replacement when $\mathfrak{g}$ is a general simple Lie algebra. By Cartan's semisimplicity criterion, mentioned in Remark 6.3.13, the Killing form

$\kappa_{\mathfrak{g}}$ is a nondegenerate $\mathfrak{g}$-invariant bilinear form on $\mathfrak{g}$, which we can use to define the Casimir operator $\Omega_{\kappa_{\mathfrak{g}},V}$ as in Theorem 6.4.2. A formula for the scalar by which this acts when V is a highest-weight module is given in [10, section 22.3], and the analogue of Proposition 7.3.2 is not hard to prove. So the proofs in which we used Casimir operators, namely those of the irreducibility of highest-weight modules (Theorem 7.3.4) and of complete reducibility (Theorem 7.5.3), go through in general. The authors mentioned above prefer to show complete reducibility first: [6] and [8] explain Weyl's original proof using compact Lie groups whereas [2] and [10] use Casimir operators, though in a slightly different way. Showing that any highest-weight module is irreducible is much easier if it is already known to be completely reducible.

The second aspect of the general story that is harder to prove than in the $\mathfrak{sl}_n$ case is the existence of an irreducible $V(\lambda)$ with a specified highest weight $\lambda \in \Lambda^+$. The tensor-product argument reduces this problem to the case of the *fundamental weights* $\varpi_1, \ldots, \varpi_\ell$ that form the basis of $\mathfrak{h}^*$ dual to $h_1, \ldots, h_\ell$. For $\mathfrak{sl}_n$, the module $V(\varpi_k)$ equals $\mathrm{Alt}^k(\mathbb{C}^n)$, by (7.1.3). For a general simple Lie algebra $\mathfrak{g}$, constructing the fundamental modules is harder.

Example 8.2.1. Let us return to the Lie algebra $\mathfrak{g} \cong \mathfrak{so}_5$ of Example 8.1.1. Easy calculations give

$$\varpi_1 = \alpha_1 + \alpha_2, \ \varpi_2 = \frac{1}{2}\alpha_1 + \alpha_2.$$

In the natural $\mathfrak{g}$-module $\mathbb{C}^5$, v_1 is a highest-weight vector of weight $\alpha_1 + \alpha_2$, and it generates the whole module:

$$f_1 v_1 = v_2, \quad f_2 f_1 v_1 = 2v_3, \quad f_2^2 f_1 v_1 = -4v_4, \quad f_1 f_2^2 f_1 v_1 = 4v_5.$$

So $\mathbb{C}^5$ is an irreducible $\mathfrak{g}$-module of highest weight ϖ_1.

Let V denote an irreducible $\mathfrak{g}$-module of highest weight ϖ_2, with $v \in V$ a highest-weight vector. We can determine a basis of V without having a prior construction of it, as follows. Since $\varpi_2(h_1) = 0$, $f_1 v = 0$. Since $\varpi_2(h_2) = 1$, $f_2^2 v = 0$ and $f_2 v$ is a weight vector of weight $\frac{1}{2}\alpha_1$. Since $\frac{1}{2}\alpha_1(h_1) = 1$, $f_1^2 f_2 v = 0$ and $f_1 f_2 v$ is a weight vector of weight $-\frac{1}{2}\alpha_1$. Since $-\frac{1}{2}\alpha_1(h_2) = 1$, $f_2^2 f_1 f_2 v = 0$ and $f_2 f_1 f_2 v$ is a weight vector of weight $-\varpi_2$. Since $-\varpi_2(h_1) = 0$, $f_1 f_2 f_1 f_2 v = 0$ and there are no other nonzero vectors of the form $f_{i_1} f_{i_2} \cdots f_{i_r} v$. Since $v, f_2 v, f_1 f_2 v, f_2 f_1 f_2 v$ have different weights, they are linearly independent and hence form a basis of V. We have seen in the course of this argument how h_1, h_2, f_1, f_2 act on each basis vector; the action of e_1, e_2 can also be determined, starting from $e_1 v = e_2 v = 0$ and using the Lie bracket relation $[e_i, f_j] = \delta_{ij} h_i$. With this knowledge in hand, it is easy to construct V, by giving explicit 4×4 matrices that represent the generators of $\mathfrak{g}$. Beware that this example is misleadingly straightforward, in that here there is no need to determine the weight multiplicities in V. ■

Fortunately, there are uniform proofs that every $V(\lambda)$ exists. The most elegant, explained in [2, chapter 10] and [10, section 20], proceeds via an infinite-dimensional *Verma module*, which is defined abstractly as satisfying the relations needed for it to be a highest-weight module of weight λ and no other relations; such a Verma module has a unique irreducible quotient, which turns out to be the desired finite-dimensional $V(\lambda)$ under the assumption that λ is a dominant integral weight. See [6, part III], [7, chapter 5], and [8, chapter 7] for other approaches to constructing $V(\lambda)$, in special cases and in general.

8.3 Characters and bases of representations

A question left unanswered in Chapter 7 was the following: what are the weight multiplicities in a general irreducible $\mathfrak{gl}_n$-module $V(\lambda)$? (The methods used in Examples 7.4.8 and 7.5.10 apply only in special cases.) There are two beautiful answers, both related to important ideas in algebraic combinatorics.

The first answer expresses the weight multiplicities of an integral $\mathfrak{gl}_n$-module V via a polynomial in the variables $x_1, x_1^{-1}, x_2, x_2^{-1}, \dots, x_n, x_n^{-1}$, known as the *character* of V,

$$\mathrm{ch}(V) = \sum_{(b_1, b_2, \dots, b_n) \in \mathbb{Z}^n} \dim V_{b_1\varepsilon_1 + b_2\varepsilon_2 + \cdots + b_n\varepsilon_n} \, x_1^{b_1} x_2^{b_2} \cdots x_n^{b_n}.$$

A similar definition can be made for modules over an arbitrary simple Lie algebra. The celebrated *Weyl character formula* determines the character of an irreducible module. In the case of $\mathfrak{gl}_n$, the statement is that, for any $a_1\varepsilon_1 + \cdots + a_n\varepsilon_n \in \Lambda^+$,

$$\mathrm{ch}\,(V(a_1\varepsilon_1 + a_2\varepsilon_2 + \cdots + a_n\varepsilon_n)) = \frac{\det((x_j^{a_i+n-i})_{1\le i,j\le n})}{\prod_{1\le i<j\le n} (x_i - x_j)}. \tag{8.3.1}$$

The numerator of the right-hand side is antisymmetric in the variables $x_1, \dots, x_n$, because interchanging two of these variables interchanges two columns of the matrix and therefore changes the sign of the determinant. It follows that the denominator does divide the numerator in the integral domain $\mathbb{Z}[x_1, x_1^{-1}, \dots, x_n, x_n^{-1}]$, so the fraction makes sense. Note that when $a_1 = \cdots = a_n = 0$ the left-hand side of (8.3.1) is 1, so in that case (8.3.1) becomes the well-known Vandermonde determinant formula

$$\det((x_j^{n-i})_{1\le i,j\le n}) = \prod_{1\le i<j\le n} (x_i - x_j). \tag{8.3.2}$$

Various proofs of the Weyl character formula are given in the books by Fulton and Harris [6], Goodman and Wallach [7], Hall [8], and Humphreys [10].

Despite the aesthetic appeal of (8.3.1), it is not immediately clear how to find the coefficient of a particular monomial $x_1^{b_1} \cdots x_n^{b_n}$ on the right-hand side. This problem has a long history in the theory of symmetric functions, for which the canonical reference is Macdonald's book [14]. We can assume without loss of generality that $a_n \geq 0$, in which case the right-hand side of (8.3.1) is a polynomial in $x_1, \ldots, x_n$ called the *Schur polynomial* $s_{(a_1,\ldots,a_n)}(x_1, \ldots, x_n)$. The coefficient of the monomial $x_1^{b_1} \cdots x_n^{b_n}$ in this polynomial equals the *Kostka number* $K_{(a_1,\ldots,a_n),(b_1,\ldots,b_n)}$, defined to be the number of *semistandard Young tableaux* of shape $(a_1, \ldots, a_n)$ and weight $(b_1, \ldots, b_n)$ [14, section I.5]. These tableaux are left-justified arrays of numbers between 1 and n, in which each row weakly decreases from left to right and each column strictly decreases from top to bottom. The shape is $(a_1, \ldots, a_n)$ if there are a_i numbers in the ith row; the weight is $(b_1, \ldots, b_n)$ if the number j occurs b_j times.

So the second answer to the weight multiplicity problem, again restricting to the case of $\mathfrak{gl}_n$, is that, for any $a_1\varepsilon_1 + \cdots + a_n\varepsilon_n \in \Lambda^+$ with $a_n \geq 0$,

$$\dim V(a_1\varepsilon_1 + \cdots + a_n\varepsilon_n)_{b_1\varepsilon_1+\cdots+b_n\varepsilon_n} = K_{(a_1,\ldots,a_n),(b_1,\ldots,b_n)}. \tag{8.3.3}$$

Example 8.3.1. To compute the weight multiplicities of the $\mathfrak{gl}_3$-module $V(3\varepsilon_1+\varepsilon_2)$ by this rule, it suffices to list the semistandard Young tableaux of shape $(3, 1, 0)$:

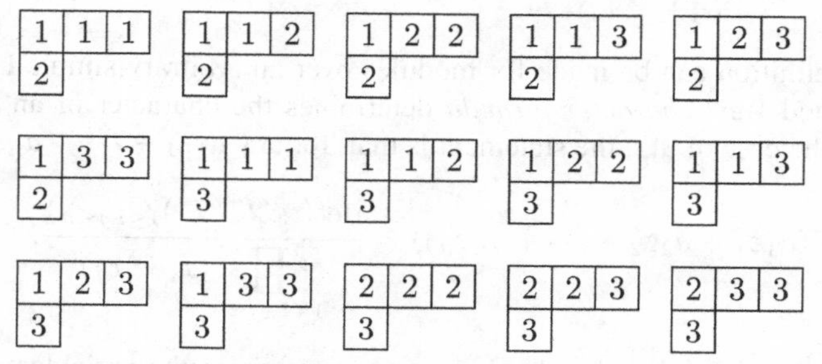

Since two of these tableaux have weight $(2, 1, 1)$, the multiplicity of $2\varepsilon_1+\varepsilon_2+\varepsilon_3$ as a weight of $V(3\varepsilon_1+\varepsilon_2)$ equals 2. Note that this agrees with what we found in Example 7.4.8, that the multiplicity of ε_1 as a weight of $V(2\varepsilon_1 - \varepsilon_3) \cong V(3\varepsilon_1 + \varepsilon_2) \otimes \mathbb{C}_{-\mathrm{tr}}$ equals 2. ■

It is natural to want to prove (8.3.3) more directly, by defining a basis $\{v_T\}$ of $V(a_1\varepsilon_1 + \cdots + a_n\varepsilon_n)$ indexed by semistandard Young tableaux T of shape $(a_1, \ldots, a_n)$, such that if T has weight $(b_1, \ldots, b_n)$ then v_T is a weight vector of weight $b_1\varepsilon_1 + \cdots + b_n\varepsilon_n$. There are (at least) three different ways to do this.

The first way, concisely described by Fulton [5, chapter 8], involves constructing $V(a_1\varepsilon_1 + \cdots + a_n\varepsilon_n)$ as the quotient of $(\mathbb{C}^n)^{\otimes(a_1+\cdots+a_n)}$ by a specific submodule, whose definition arises naturally from the representation theory of the symmetric group. This forms part of a general description of the irreducible $\mathfrak{gl}_n$-submodules of $(\mathbb{C}^n)^{\otimes k}$ in terms of the irreducible representations of S_k, known as *Schur–Weyl duality*. See Goodman and Wallach [7, chapter 9] for the full story.

The second way, producing what is called the *Gel'fand–Tsetlin basis*, arises by regarding an irreducible $\mathfrak{gl}_n$-module as a $\mathfrak{gl}_{n-1}$-module, where $\mathfrak{gl}_{n-1}$ is identified with the subalgebra of $\mathfrak{gl}_n$ consisting of matrices whose bottom row and last column are zero. There is an easy *branching rule* to determine which irreducible $\mathfrak{gl}_{n-1}$-modules occur in this restriction [7, chapter 8]; it turns out that they all occur with multiplicity 1, so they are uniquely determined as submodules. This allows a recursive construction: given a basis for each irreducible $\mathfrak{gl}_{n-1}$-module, we obtain a basis for each irreducible $\mathfrak{gl}_n$-module. Both Schur–Weyl duality and the branching rule have analogues for other matrix Lie algebras such as $\mathfrak{so}_n$; these are treated thoroughly in [7].

The third way is more difficult but applies in the context of a general simple Lie algebra. We know from Proposition 7.2.11 that an irreducible $V(\lambda)$ is spanned by weight vectors of the form $f_{i_1} f_{i_2} \cdots f_{i_r} v$, where v is a fixed highest-weight vector. In various special cases we found very natural choices of linear combinations of these spanning vectors which formed a basis of $V(\lambda)$. For instance, for the $\mathfrak{sl}_2$-module $V(m)$ in Chapter 5, the string basis vectors $\frac{1}{i!} f^i v$ were an obvious choice; for the fundamental representation $V(\varpi_2)$ of $\mathfrak{so}_5$ in Example 8.2.1, the basis $v, f_2 v, f_1 f_2 v, f_2 f_1 f_2 v$ arose naturally. The 'obvious' bases of $\mathrm{Sym}^k(\mathbb{C}^n)$ and $\mathrm{Alt}^k(\mathbb{C}^n)$ were expressed as linear combinations of weight vectors $f_{i_1} f_{i_2} \cdots f_{i_r} v$ in (7.4.2) and (7.4.4), but in those cases the expression was apparently more arbitrary.

It would be natural to presume that the form of a basis of $V(\lambda)$ would depend heavily on λ. In fact, however, once a simple Lie algebra $\mathfrak{g}$ is fixed there is a uniform way to specify a basis of every irreducible $\mathfrak{g}$-module simultaneously. Suppose $\mathfrak{g} = \mathfrak{sl}_3$, and consider the expressions

$$\frac{1}{i!j!k!} f_1^i f_2^j f_1^k v \quad \text{and} \quad \frac{1}{i!j!k!} f_2^i f_1^j f_2^k v$$
$$\text{for all integers } i, j, k \geq 0 \quad \text{with} \quad i + k \leq j.$$

For any particular irreducible $\mathfrak{sl}_3$-module, if v is taken to be a highest-weight vector for that $\mathfrak{sl}_3$-module then most of these expressions will be zero. Amazingly, the nonzero ones always form a basis, called the *canonical basis* or *global crystal basis*. Furthermore, there is an analogous set of expressions for any simple Lie algebra $\mathfrak{g}$; in general they are linear combinations of expressions of the form $f_{i_1} \cdots f_{i_r} v$, not simply multiples of them.

The original proofs of this result, found by Lusztig and Kashiwara, both make use of the *quantized universal enveloping algebra* of $\mathfrak{g}$. This is the most advanced topic we have mentioned. The ambitious reader, who has already mastered the concept of the *universal enveloping algebra* as defined in the book by Humphreys [10, section 17], is advised to try the texts of Hong and Kang [9] or Jantzen [11]; the former includes an explanation of how the combinatorics of crystal bases reduces to that of Young tableaux in the $\mathfrak{sl}_n$ case. Thus crystal bases can be seen to provide the true generalization of (8.3.3) to arbitrary simple Lie algebras.

APPENDIX

Solutions to the exercises

Solutions for Chapter 2 exercises

Solution to Exercise 2.3.1. We start from the Jacobi identity and substitute the given formulas, using the bilinearity and skew-symmetry of the Lie bracket to simplify the results:

$$\begin{aligned}
[x_1, [x_2, x_3]] &= [[x_1, x_2], x_3] + [x_2, [x_1, x_3]], \\
[x_1, x_2 + bx_3] &= [x_1 + x_2, x_3] + [x_2, ax_1 + x_3], \\
[x_1, x_2] + b[x_1, x_3] &= [x_1, x_3] + [x_2, x_3] + a[x_2, x_1] + [x_2, x_3], \\
x_1 + x_2 + b(ax_1 + x_3) &= ax_1 + x_3 + x_2 + bx_3 - a(x_1 + x_2) + x_2 + bx_3, \\
(ab + 1)x_1 + x_2 + bx_3 &= (2 - a)x_2 + (2b + 1)x_3.
\end{aligned}$$

Since x_1, x_2, x_3 are linearly independent, we can conclude that

$$ab + 1 = 0,\ 1 = 2 - a,\ b = 2b + 1,$$

which forces $a = 1$, $b = -1$.

If you think about it further, this calculation suffices to ensure the existence of the Lie algebra described in the question with these values of a and b, namely $\mathfrak{g}$ with basis x_1, x_2, x_3 whose Lie bracket satisfies

$$[x_1, x_2] = x_1 + x_2, \quad [x_1, x_3] = x_1 + x_3, \quad [x_2, x_3] = x_2 - x_3.$$

It is clear that subject to these equations, there is a unique way to define the bracket on linear combinations of x_1, x_2, x_3 so that it will be bilinear and skew-symmetric; what we are claiming is that this bracket then satisfies the Jacobi identity. It suffices to check this when the three elements involved belong to $\{x_1, x_2, x_3\}$, and the identity holds automatically when two of the three elements coincide, so we are reduced to the case where the three elements involved are x_1, x_2, x_3 in some particular order. The above calculation (in which the steps were all reversible) demonstrates the Jacobi identity when the elements involved are x_1, x_2, x_3 in that order, and easy manipulations show that the identities for the other five possible orders are equivalent.

Solution to Exercise 2.3.2. It is clear that $[\cdot,\cdot]$ is bilinear and skew-symmetric if and only if the function f is bilinear and skew-symmetric. Assuming this to be the case, the Jacobi identity can be simplified successively into the following equivalent forms, which must hold for all $x, y, z \in \mathfrak{g}$:

$$\begin{aligned}
[x,[y,z]] &= [[x,y],z] + [y,[x,z]],\\
[x, f(y,z)v] &= [f(x,y)v, z] + [y, f(x,z)v],\\
f(y,z)[x,v] &= f(x,y)[v,z] + f(x,z)[y,v],\\
f(y,z)f(x,v)v &= f(x,y)f(v,z)v + f(x,z)f(y,v)v,\\
f(y,z)f(x,v) &= f(x,y)f(v,z) + f(x,z)f(y,v).
\end{aligned}$$

If $f(x,v) = 0$ for all $x \in \mathfrak{g}$ then this last equality certainly holds. Otherwise, the linear function $\alpha : \mathfrak{g} \to \mathbb{C}$ defined by $\alpha(x) = f(x,v)$ is nonzero, so there is some $v' \in \mathfrak{g}$ such that $\alpha(v') = 1$. Note that v and v' are linearly independent, since $\alpha(v) = 0$. Define another linear function $\beta : \mathfrak{g} \to \mathbb{C}$ by $\beta(x) = f(v', x)$. Then $\beta(v) = 1$ and $\beta(v') = 0$. Setting $x = v'$ in the above equation gives

$$f(y,z) = \alpha(y)\beta(z) - \beta(y)\alpha(z) \text{ for all } x, y \in \mathfrak{g}.$$

Conversely, if f is defined in terms of linear functions α and β in this way then it is obviously bilinear and skew-symmetric; assuming that $\alpha(v) = 0$ and $\beta(v) = 1$, we have that $f(x,v) = \alpha(x)$ and that f also satisfies a condition equivalent to the Jacobi identity, because

$$\begin{aligned}
f(y,z)f(x,v) &= \Big(\alpha(y)\beta(z) - \beta(y)\alpha(z)\Big)\alpha(x)\\
&= -\Big(\alpha(x)\beta(y) - \beta(x)\alpha(y)\Big)\alpha(z) + \Big(\alpha(x)\beta(z) - \beta(x)\alpha(z)\Big)\alpha(y)\\
&= f(x,y)f(v,z) + f(x,z)f(y,v).
\end{aligned}$$

So the answer is that *either* f is a skew-symmetric bilinear form on $\mathfrak{g}$ and v is such that $f(x,v) = 0$ for all $x \in \mathfrak{g}$ *or* there are two linear functions $\alpha, \beta : \mathfrak{g} \to \mathbb{C}$ such that $\alpha(v) = 0$, $\beta(v) = 1$ and $f : \mathfrak{g} \times \mathfrak{g} \to \mathbb{C}$ is defined in terms of α and β as above.

Solution to Exercise 2.3.3. If $\mathfrak{g}$ is an abelian Lie algebra and $\mathfrak{g}'$ is an arbitrary Lie algebra then, by definition, a homomorphism $\varphi : \mathfrak{g}' \to \mathfrak{g}$ is a linear map such that

$$\varphi([x,y]) = 0 \quad \text{for all } x, y \in \mathfrak{g}'.$$

If $\mathfrak{g}' = \mathfrak{sl}_2$, this condition forces

$$\varphi(h) = \varphi([e,f]) = 0, \quad \varphi(2e) = \varphi([h,e]) = 0, \quad \varphi(-2f) = \varphi([h,f]) = 0,$$

which by linearity forces φ to be the zero map, as claimed.

If $\mathfrak{g}$ is an abelian Lie algebra and $\mathfrak{g}'$ is an arbitrary Lie algebra then, by definition, a homomorphism $\varphi : \mathfrak{g} \to \mathfrak{g}'$ is a linear map such that

$$0 = [\varphi(x), \varphi(y)] \quad \text{for all } x, y \in \mathfrak{g}.$$

For example, we can define a nonzero homomorphism φ from the abelian Lie algebra $\mathbb{C}$ to $\mathfrak{sl}_2$ by the rule $\varphi(a) = ae$ (where e could be replaced by any nonzero element of $\mathfrak{sl}_2$).

Solution to Exercise 2.3.4. If I is a subset of $\{e_{11}, e_{12}, e_{21}, e_{22}\}$ then the span of I has dimension $|I|$. For the purposes of classification we need only compare subsets of the same cardinality. We will treat the different possible values of $|I|$ in turn.

The span of the subset $I = \emptyset$, namely the zero subspace $\{0\}$, is obviously closed under taking commutators.

If $|I| = 1$ then the span of I is certainly closed under the taking of commutators; in fact, any two elements commute because one is a scalar multiple of the other. These one-dimensional Lie algebras $\mathbb{C}e_{11}, \mathbb{C}e_{12}, \mathbb{C}e_{21}, \mathbb{C}e_{22}$ are all abelian and isomorphic to $\mathbb{C}$.

If $|I| = 2$ then the span of I is closed under the taking of commutators if and only if the commutator of the two elements of I is a linear combination of those two elements. As seen in Example 2.2.3, this holds for $I = \{e_{11}, e_{22}\}$ because $[e_{11}, e_{22}] = 0$, resulting in the two-dimensional abelian Lie algebra $\mathbb{C}\{e_{11}, e_{22}\}$. It also holds for $I = \{e_{11}, e_{12}\}$ because $[e_{11}, e_{12}] = e_{12}$, resulting in the two-dimensional non-abelian Lie algebra $\mathbb{C}\{e_{11}, e_{12}\}$. To investigate the other cases, we calculate

$$\begin{aligned}
[e_{11}, e_{21}] &= -e_{21}, \\
[e_{12}, e_{21}] &= e_{11} - e_{22}, \\
[e_{12}, e_{22}] &= e_{12}, \\
[e_{21}, e_{22}] &= -e_{21}.
\end{aligned}$$

The case $I = \{e_{12}, e_{21}\}$ is the only case where the closure condition does not hold. So we get three further two-dimensional Lie algebras, $\mathbb{C}\{e_{11}, e_{21}\}$, $\mathbb{C}\{e_{12}, e_{22}\}$, and $\mathbb{C}\{e_{21}, e_{22}\}$. These are all non-abelian so, by the classification of two-dimensional Lie algebras, they are isomorphic to each other and to $\mathbb{C}\{e_{11}, e_{12}\}$.

If $|I| = 3$, there are four possibilities for I. If either e_{11} or e_{22} is the basis element omitted from I then the above calculation of $[e_{12}, e_{21}]$ shows that the span of I is not closed under taking commutators. The other two possibilities give three-dimensional Lie algebras $\mathbb{C}\{e_{11}, e_{12}, e_{22}\}$ and $\mathbb{C}\{e_{11}, e_{21}, e_{22}\}$. There is an isomorphism between these Lie algebras that simply replaces every 1 in a subscript with a 2 and vice versa:

$$\begin{aligned}
\mathbb{C}\{e_{11}, e_{12}, e_{22}\} &\xrightarrow{\sim} \mathbb{C}\{e_{11}, e_{21}, e_{22}\}, \\
e_{11} &\mapsto e_{22}, \\
e_{12} &\mapsto e_{21}, \\
e_{22} &\mapsto e_{11}.
\end{aligned}$$

This map is clearly an isomorphism of vector spaces. Saying that it is an isomorphism of Lie algebras reduces to saying that the commutators of corresponding basis elements also correspond, which is true by the following calculation:

$$\begin{aligned} [e_{11}, e_{12}] &= e_{12} \mapsto e_{21} = [e_{22}, e_{21}], \\ [e_{11}, e_{22}] &= 0 \mapsto 0 = [e_{22}, e_{11}], \\ [e_{12}, e_{22}] &= e_{12} \mapsto e_{21} = [e_{21}, e_{11}]. \end{aligned}$$

Finally, if $I = \{e_{11}, e_{12}, e_{21}, e_{22}\}$ then the span of I is the whole of $\mathfrak{gl}_2$, which is obviously closed under taking commutators.

Solution to Exercise 2.3.5. For part (i), direct matrix calculations give $[x, y] = z$, $[y, z] = x$, and $[z, x] = y$. (In other words, $\mathfrak{so}_3$ with its Lie bracket is just a complex version of $\mathbb{R}^3$ with the vector cross product.) For part (ii) we must find a basis $\varphi(e), \varphi(h), \varphi(f)$ of $\mathfrak{so}_3$ that satisfies the $\mathfrak{sl}_2$ basis relations. As the hint suggests, we will try for a solution where $\varphi(h) = dx$ for some $d \in \mathbb{C}^\times$. If $\varphi(e) = ax + by + cz$ and $\varphi(f) = a'x + b'y + c'z$ then the required relations are

$$\begin{aligned} dx &= [ax + by + cz, a'x + b'y + c'z] \\ &= (bc' - cb')x + (ca' - ac')y + (ab' - ba')z, \\ 2ax + 2by + 2cz &= [dx, ax + by + cz] = -cdy + bdz, \\ -2a'x - 2b'y - 2c'z &= [dx, a'x + b'y + c'z] = -c'dy + b'dz. \end{aligned}$$

Equating coefficients, we find that $a = a' = 0$, $2b = -cd$, $2c = bd$, $2b' = c'd$, $2c' = -b'd$, and $bc' - cb' = d$. Since b and c cannot both be zero (or else $\varphi(e)$ and $\varphi(h)$ would be linearly dependent), we deduce from $2b = -cd$ and $2c = bd$ that $d = \pm 2\mathbf{i}$. Selecting $d = 2\mathbf{i}$, we can easily arrange for all the other equations to be satisfied: one solution is

$$\varphi(h) = 2\mathbf{i}x, \quad \varphi(e) = y + \mathbf{i}z, \quad \varphi(f) = -y + \mathbf{i}z.$$

Non-real numbers are necessarily involved here, since $\mathfrak{so}_3(\mathbb{R}) \not\cong \mathfrak{sl}_2(\mathbb{R})$.

For part (iii), we calculate first that

$$\psi(e) = \begin{pmatrix} 0 & 2 & 0 \\ 0 & 0 & 1 \\ 0 & 0 & 0 \end{pmatrix}, \quad \psi(h) = \begin{pmatrix} 2 & 0 & 0 \\ 0 & 0 & 0 \\ 0 & 0 & -2 \end{pmatrix}, \quad \psi(f) = \begin{pmatrix} 0 & 0 & 0 \\ 1 & 0 & 0 \\ 0 & 2 & 0 \end{pmatrix}.$$

It suffices to find $g \in GL_3$ satisfying the three equations

$$g\psi(e) = \varphi(e)g, \quad g\psi(h) = \varphi(h)g, \quad g\psi(f) = \varphi(f)g.$$

Since the standard basis vectors v_1, v_2, v_3 of $\mathbb{C}^3$ are eigenvectors of $\psi(h)$ with eigenvalues $2, 0, -2$ respectively, the second equation, $g\psi(h) = \varphi(h)g$, implies (actually, is equivalent to saying) that the columns gv_1, gv_2, gv_3 of g are eigenvectors of

$\varphi(h) = 2\mathrm{i}x$ with these same eigenvalues. We can easily verify that the eigenvalues of $2\mathrm{i}x$ are indeed $2, 0, -2$. Finding the corresponding eigenvectors by basic linear algebra, we see that

$$g = \begin{pmatrix} a\mathrm{i} & 0 & c \\ a & 0 & c\mathrm{i} \\ 0 & b & 0 \end{pmatrix} \quad \text{for some } a, b, c \in \mathbb{C}^\times.$$

Now the first equation, $g\psi(e) = \varphi(e)g$, becomes

$$\begin{pmatrix} a\mathrm{i} & 0 & c \\ a & 0 & c\mathrm{i} \\ 0 & b & 0 \end{pmatrix} \begin{pmatrix} 0 & 2 & 0 \\ 0 & 0 & 1 \\ 0 & 0 & 0 \end{pmatrix} = \begin{pmatrix} 0 & 0 & \mathrm{i} \\ 0 & 0 & 1 \\ -\mathrm{i} & -1 & 0 \end{pmatrix} \begin{pmatrix} a\mathrm{i} & 0 & c \\ a & 0 & c\mathrm{i} \\ 0 & b & 0 \end{pmatrix},$$

which reduces to $b = 2a$, $c = a\mathrm{i}$. Since multiplying g by a scalar does not affect the map of conjugation by g, we may as well put $a = 1$, so that

$$g = \begin{pmatrix} \mathrm{i} & 0 & \mathrm{i} \\ 1 & 0 & -1 \\ 0 & 2 & 0 \end{pmatrix},$$

which is easily seen to satisfy the third equation as well.

Solution to Exercise 2.3.6. Only two three-dimensional Lie algebras appeared in Exercise 2.3.4, and they were isomorphic to each other. So it suffices to prove that one of them, say $\mathbb{C}\{e_{11}, e_{12}, e_{22}\}$, is not isomorphic to $\mathfrak{sl}_2$. Suppose for a contradiction that there is some Lie algebra isomorphism $\varphi : \mathbb{C}\{e_{11}, e_{12}, e_{22}\} \xrightarrow{\sim} \mathfrak{sl}_2$. Then $\varphi(e_{11}), \varphi(e_{12}), \varphi(e_{22})$ form a basis of $\mathfrak{sl}_2$ and satisfy the same Lie bracket relations as the elements e_{11}, e_{12}, e_{22}, namely

$$[\varphi(e_{11}), \varphi(e_{12})] = \varphi(e_{12}), \quad [\varphi(e_{11}), \varphi(e_{22})] = 0, \quad [\varphi(e_{12}), \varphi(e_{22})] = \varphi(e_{12}).$$

By the bilinearity and skew-symmetry of the Lie bracket, these equations imply that $[x, y] \in \mathbb{C}\varphi(e_{12})$ for all $x, y \in \mathfrak{sl}_2$. Since $[e, f], [h, e], [h, f]$ are linearly independent, this is clearly false.

Solution to Exercise 2.3.7. For part (i), any linear transformation τ of $\mathfrak{l}_2$ has some matrix $\left(\begin{smallmatrix} a & b \\ c & d \end{smallmatrix}\right)$ with respect to the standard basis s, t; τ is invertible if and only if $ad - bc \neq 0$. As we have seen, τ respects the Lie bracket if and only if $\tau([s, t]) = [\tau(s), \tau(t)]$, which is equivalent to saying that

$$bs + dt = [as + ct, bs + dt] = (ad - bc)t, \quad \text{i.e. } b = 0,\ d = ad.$$

Hence τ is an automorphism of $\mathfrak{l}_2$ if and only if $a = 1$, $b = 0$, and $d \neq 0$: we then have

$$\tau(s) = s + ct, \quad \tau(t) = dt.$$

For part (ii), the conjugation map $x \mapsto gxg^{-1}$ is certainly a linear map from $\mathfrak{gl}_n$ to itself and its inverse is $x \mapsto g^{-1}xg$, so we need only show that it respects commutators, which is a simple calculation:

$$\begin{aligned}[gxg^{-1}, gyg^{-1}] &= gxg^{-1}gyg^{-1} - gyg^{-1}gxg^{-1} \\ &= gxyg^{-1} - gyxg^{-1} = g[x,y]g^{-1}.\end{aligned}$$

Similarly, the only non-obvious part of (iii) is that the map respects commutators, which holds because

$$\begin{aligned}[-x^{\mathsf{t}}, -y^{\mathsf{t}}] &= (-x^{\mathsf{t}})(-y^{\mathsf{t}}) - (-y^{\mathsf{t}})(-x^{\mathsf{t}}) \\ &= x^{\mathsf{t}}y^{\mathsf{t}} - y^{\mathsf{t}}x^{\mathsf{t}} \\ &= (yx)^{\mathsf{t}} - (xy)^{\mathsf{t}} = -[x,y]^{\mathsf{t}}.\end{aligned}$$

In part (iv), note first that every conjugation automorphism $x \mapsto gxg^{-1}$ of $\mathfrak{gl}_2$, where $g \in GL_2$, does restrict to an automorphism of $\mathfrak{sl}_2$ because conjugating a matrix leaves its trace unchanged. We want to prove that every automorphism τ of $\mathfrak{sl}_2$ is of this form. As suggested by the hint, we are allowed to replace τ by the automorphism $x \mapsto g'\tau(x)(g')^{-1}$ for any $g' \in GL_2$. By the Jordan-form theorem, there is some g' such that $g'\tau(h)(g')^{-1}$ is in Jordan form, so we can assume that $\tau(h)$ is itself in Jordan form. Now the 2×2 Jordan-form matrices with trace zero are

$$\begin{pmatrix} 0 & 0 \\ 0 & 0 \end{pmatrix} = 0, \quad \begin{pmatrix} 0 & 1 \\ 0 & 0 \end{pmatrix} = e, \quad \begin{pmatrix} d & 0 \\ 0 & -d \end{pmatrix} = dh \text{ for } d \in \mathbb{C}^\times.$$

Since τ is injective, we know that $\tau(h) \neq 0$. Suppose that $\tau(h) = e$. Then, if $\tau(e) = ae + bh + cf$, we find

$$2ae + 2bh + 2cf = \tau([h,e]) = [\tau(h), \tau(e)] = [e, ae + bh + cf] = -2be + ch,$$

from which we deduce that $a = b = c = 0$, contradicting $\tau(e) \neq 0$. So it must be that $\tau(h) = dh$ for some $d \in \mathbb{C}^\times$. Now if $\tau(e) = ae+bh+cf$ and $\tau(f) = a'e+b'h+c'f$, we have

$$\begin{aligned}2ae + 2bh + 2cf &= [\tau(h), \tau(e)] = [dh, ae + bh + cf] = 2ade - 2cdf, \\ -2a'e - 2b'h - 2c'f &= [\tau(h), \tau(f)] = [dh, a'e + b'h + c'f] = 2a'de - 2c'df.\end{aligned}$$

Equating coefficients, we find that $b = b' = 0$, and either $d = 1$, $c = a' = 0$ or $d = -1$, $a = c' = 0$. Using the final fact that $\tau([e,f]) = [\tau(e), \tau(f)]$, we then deduce that τ is either

$$e \mapsto ae, \quad h \mapsto h, \quad f \mapsto a^{-1}f,$$

which is conjugation by $\left(\begin{smallmatrix} a & 0 \\ 0 & 1 \end{smallmatrix}\right)$, or

$$e \mapsto cf, \quad h \mapsto -h, \quad f \mapsto c^{-1}e,$$

which is conjugation by $\left(\begin{smallmatrix} 0 & 1 \\ c & 0 \end{smallmatrix}\right)$. (Note that the restriction to $\mathfrak{sl}_2$ of the automorphism in part (iii) is the case $c = -1$ of the latter.) So we have proved that every automorphism of $\mathfrak{sl}_2$ is conjugation by some element of GL_2.

Solution to Exercise 2.3.8. For (i), it suffices to show that the Jacobi identity holds when the three elements involved belong to $\{x, y, z\}$. In fact, as observed in Exercise 2.3.1, it suffices to prove the single equation

$$[[x, y], z] + [[y, z], z] + [[z, x], y] = 0.$$

But from the definition of the Lie bracket in $\mathfrak{g}_{\left(\begin{smallmatrix} a & b \\ c & d \end{smallmatrix}\right)}$, all three terms on the left-hand side vanish.

For part (ii), it is convenient to identify $\mathfrak{g}_{\left(\begin{smallmatrix} a & b \\ c & d \end{smallmatrix}\right)}$ and $\mathfrak{g}_{\left(\begin{smallmatrix} a' & b' \\ c' & d' \end{smallmatrix}\right)}$ as vector spaces, i.e. treat x, y, z as the basis of both; we must then distinguish the two Lie brackets, so call them $[\cdot, \cdot]$ and $[\cdot, \cdot]'$ respectively. The two Lie algebras are isomorphic if and only if there is some invertible linear transformation ψ of $\mathbb{C}\{x, y, z\}$ such that

$$\psi([x_1, x_2]) = [\psi(x_1), \psi(x_2)]' \text{ for all } x_1, x_2 \in \mathbb{C}\{x, y, z\}.$$

For the usual reasons it is equivalent to require the three specific equations

$$\psi([x, y]) = [\psi(x), \psi(y)]', \; \psi([x, z]) = [\psi(x), \psi(z)]', \; \psi([y, z]) = [\psi(y), \psi(z)]'.$$

Now since $\left(\begin{smallmatrix} a & b \\ c & d \end{smallmatrix}\right)$ is invertible, $ay+cz$ and $by+dz$ form a basis of $\mathbb{C}\{y, z\}$, so $\mathbb{C}\{y, z\}$ is the span of all Lie brackets of $\mathfrak{g}_{\left(\begin{smallmatrix} a & b \\ c & d \end{smallmatrix}\right)}$, and similarly of $\mathfrak{g}_{\left(\begin{smallmatrix} a' & b' \\ c' & d' \end{smallmatrix}\right)}$ also. Therefore the putative ψ must preserve $\mathbb{C}\{y, z\}$, which means that

$$[\psi(y), \psi(z)]' = 0 = \psi(0) = \psi([y, z]),$$

guaranteeing the third equation. The first two equations can be encapsulated in a single equation of linear transformations of $\mathbb{C}\{y, z\}$, namely $\psi|_{\mathbb{C}\{y,z\}} \circ \varphi = \varphi' \circ \psi|_{\mathbb{C}\{y,z\}}$, where

$$\varphi(w) = [x, w], \; \varphi'(w) = [\psi(x), w]', \text{ for all } w \in \mathbb{C}\{y, z\}.$$

We now translate this into an equation of matrices, using the basis y, z. The matrix of φ is $\left(\begin{smallmatrix} a & b \\ c & d \end{smallmatrix}\right)$ by definition. The invertibility of ψ is equivalent to the invertibility of $\psi|_{\mathbb{C}\{y,z\}}$ together with the fact that the coefficient of x in $\psi(x)$ is nonzero. If the latter coefficient is α say, the matrix of φ' is clearly $\alpha\left(\begin{smallmatrix} a' & b' \\ c' & d' \end{smallmatrix}\right)$. So the equation of matrices is $g\left(\begin{smallmatrix} a & b \\ c & d \end{smallmatrix}\right) = \alpha\left(\begin{smallmatrix} a' & b' \\ c' & d' \end{smallmatrix}\right)g$, where g is the matrix of the invertible linear transformation $\psi|_{\mathbb{C}\{y,z\}}$. The result stated in (ii) follows by setting $\lambda = \alpha^{-1}$.

Part (iii) is now a pure linear algebra problem, of classifying invertible 2×2 matrices up to the equivalence relation $\sim$ defined by

$$x \sim y \Longleftrightarrow x = \lambda g y g^{-1}, \text{ for some } \lambda \in \mathbb{C}^\times, g \in GL_2.$$

Without the scalar multiplication, this would be exactly the problem solved by the Jordan form theorem; so it suffices to classify matrices in Jordan form up to $\sim$. Clearly the invertible 2×2 Jordan blocks are all $\sim \left(\begin{smallmatrix} 1 & 1 \\ 0 & 1 \end{smallmatrix}\right)$, so all that remains is the diagonal matrices, where

$$\begin{pmatrix} \alpha_1 & 0 \\ 0 & \alpha_2 \end{pmatrix} \sim \begin{pmatrix} \alpha_1' & 0 \\ 0 & \alpha_2' \end{pmatrix} \iff \{\alpha_1, \alpha_2\} = \{\lambda\alpha_1', \lambda\alpha_2'\} \text{ for some } \lambda \in \mathbb{C}^\times.$$

Clearly any diagonal matrix is equivalent to one of the form $\left(\begin{smallmatrix} \alpha & 0 \\ 0 & 1 \end{smallmatrix}\right)$, and α and α^{-1} give the same equivalence class, but there are no other equivalences. This is what the statement of (iii) says.

To answer the final question about the Lie algebra $\mathfrak{g}$ from Exercise 2.3.1, we try to find an isomorphism $\varphi : \mathfrak{g}_{\left(\begin{smallmatrix} a & b \\ c & d \end{smallmatrix}\right)} \xrightarrow{\sim} \mathfrak{g}$ for some $\left(\begin{smallmatrix} a & b \\ c & d \end{smallmatrix}\right) \in GL_2$. Considering the span of the Lie brackets, we see that we will have to have $\varphi(\mathbb{C}\{y, z\}) = \mathbb{C}\{x_1+x_2, x_1+x_3\}$, so we try setting $\varphi(y) = x_1 + x_2$ and $\varphi(z) = x_1 + x_3$, with the arbitrary choice $\varphi(x) = x_1$. Certainly $x_1, x_1 + x_2, x_1 + x_3$ are linearly independent, so this does define a vector space isomorphism φ. For φ to preserve the Lie bracket, we must have

$$\begin{aligned} a(x_1 + x_2) + c(x_1 + x_3) &= [\varphi(x), \varphi(y)] = [x_1, x_1 + x_2] = x_1 + x_2, \\ b(x_1 + x_2) + d(x_1 + x_3) &= [\varphi(x), \varphi(z)] = [x_1, x_1 + x_3] = x_1 + x_3, \\ 0 &= [\varphi(y), \varphi(z)] = [x_1 + x_2, x_1 + x_3] = 0. \end{aligned}$$

The last equation is automatic, and the first two force $a = d = 1$, $b = c = 0$. So $\mathfrak{g}$ does indeed fit into our classification: it is isomorphic to $\mathfrak{g}_{\left(\begin{smallmatrix} 1 & 0 \\ 0 & 1 \end{smallmatrix}\right)}$.

Solutions for Chapter 3 exercises

Solution to Exercise 3.4.1. Let $\mathfrak{h}$ be the subalgebra of $\mathfrak{gl}_2$ generated by x and e_{11}. In addition to these two elements, $\mathfrak{h}$ must contain

$$[x, e_{11}] = -be_{12} + ce_{21} \quad \text{and} \quad [[x, e_{11}], e_{11}] = be_{12} + ce_{21}.$$

If $bc \neq 0$ then $\mathfrak{h}$ contains e_{12} and e_{21} and hence also $[e_{12}, e_{21}] = e_{11} - e_{22}$. Therefore $\mathfrak{h}$ contains a basis of $\mathfrak{gl}_2$, so $\mathfrak{h} = \mathfrak{gl}_2$, proving (i).

For (ii), we need to consider various cases. Recall that in Exercise 2.3.4 we determined which spans of standard basis elements of $\mathfrak{gl}_2$ are subalgebras.

- If $b = c = d = 0$ then $x \in \mathbb{C}e_{11}$, so $\mathfrak{h} = \mathbb{C}e_{11}$.
- If $b = c = 0$ and $d \neq 0$ then the span of x and e_{11} is $\mathbb{C}\{e_{11}, e_{22}\}$, which is a subalgebra, so $\mathfrak{h} = \mathbb{C}\{e_{11}, e_{22}\}$.
- If $b = d = 0$ and $c \neq 0$ then the span of x and e_{11} is $\mathbb{C}\{e_{11}, e_{21}\}$, which is a subalgebra, so $\mathfrak{h} = \mathbb{C}\{e_{11}, e_{21}\}$.

- Similarly, if $c = d = 0$ and $b \neq 0$ then $\mathfrak{h} = \mathbb{C}\{e_{11}, e_{12}\}$.
- If $b = 0$ and $c, d \neq 0$ then $\mathfrak{h}$ contains e_{21} by the above calculation. The span of x, e_{11}, and e_{21} is $\mathbb{C}\{e_{11}, e_{21}, e_{22}\}$, which is a subalgebra, so $\mathfrak{h} = \mathbb{C}\{e_{11}, e_{21}, e_{22}\}$.
- Similarly, if $c = 0$ and $b, d \neq 0$ then $\mathfrak{h} = \mathbb{C}\{e_{11}, e_{12}, e_{22}\}$.

Solution to Exercise 3.4.2. To prove (i), since $\mathrm{Der}(A)$ and $\mathrm{Der}_{\mathrm{Lie}}(A)$ are Lie subalgebras of $\mathfrak{gl}(A)$ we need only show that $\mathrm{Der}(A)$ is contained in $\mathrm{Der}_{\mathrm{Lie}}(A)$. This is easy: if $\varphi \in \mathrm{Der}(A)$ then for any $a, b \in A$ we have

$$\begin{aligned}\varphi([a, b]) &= \varphi(ab - ba) = \varphi(ab) - \varphi(ba) \\ &= \varphi(a)b + a\varphi(b) - \varphi(b)a - b\varphi(a) \\ &= [\varphi(a), b] + [a, \varphi(b)],\end{aligned}$$

showing that $\varphi \in \mathrm{Der}_{\mathrm{Lie}}(A)$.

For (ii), we can take A to be $\mathbb{C}[x]/(x^2)$, in other words the two-dimensional algebra with basis $1, x$ and multiplication defined by the rule $x^2 = 0$. Since A is commutative, $\mathrm{Der}_{\mathrm{Lie}}(A)$ is the whole of $\mathfrak{gl}(A)$. The linear transformation $\varphi : A \to A$ defined by $\varphi(a + bx) = bx$ lies in $\mathrm{Der}(A)$, since

$$\begin{aligned}\varphi((a + bx)(c + dx)) &= \varphi(ac + (ad + bc)x) \\ &= (ad + bc)x \\ &= bx(c + dx) + (a + bx)dx \\ &= \varphi(a + bx)(c + dx) + (a + bx)\varphi(c + dx)\end{aligned}$$

for all $a, b, c, d \in \mathbb{C}$. (In fact, φ is the derivation of A induced by the derivation $x(d/dx)$ of $\mathbb{C}[x]$.) If we define another linear transformation $\psi : A \to A$ by $\psi(a + bx) = ax$ then

$$[\varphi, \psi](a + bx) = \varphi(\psi(a + bx)) - \psi(\varphi(a + bx)) = \varphi(ax) - \psi(bx) = ax - 0 = ax.$$

Thus $[\varphi, \psi] = \psi$, which does not belong to $\mathrm{Der}(A)$ (consider applying it to 1×1). So $\mathrm{Der}(A)$ is not an ideal of $\mathrm{Der}_{\mathrm{Lie}}(A)$.

Solution to Exercise 3.4.3. Let $\ell = \dim \mathfrak{h}$, and choose a basis $x_1, \ldots, x_\ell$ of $\mathfrak{h}$. We can then find $y_1, \ldots, y_k \in \mathfrak{g}$ such that $x_1, \ldots, x_\ell, y_1, \ldots, y_k$ is a basis of $\mathfrak{g}$. The derived algebra $\mathcal{D}\mathfrak{g}$ is spanned by the Lie brackets $[x_i, x_j]$ for $1 \leq i < j \leq \ell$, $[x_i, y_j]$ for $1 \leq i \leq \ell$ and $1 \leq j \leq k$, and $[y_i, y_j]$ for $1 \leq i < j \leq k$. Since $\mathfrak{h}$ is an ideal, $[x_i, x_j] \in \mathfrak{h}$ and $[x_i, y_j] \in \mathfrak{h}$. Hence $\mathcal{D}\mathfrak{g}$ is contained in $\mathfrak{h} + \mathbb{C}\{[y_i, y_j] \mid 1 \leq i < j \leq k\}$, showing that $\dim \mathcal{D}\mathfrak{g} \leq \ell + \binom{k}{2}$. If $k = 1$ or $k = 2$ then $\ell + \binom{k}{2} < \ell + k = \dim \mathfrak{g}$, proving (i).

If $\mathfrak{h}$ is the centre $Z(\mathfrak{g})$ then the Lie brackets $[x_i, x_j]$ and $[x_i, y_j]$ are zero so that $\mathcal{D}\mathfrak{g} = \mathbb{C}\{[y_i, y_j] \mid 1 \leq i < j \leq k\}$, showing that $\dim \mathcal{D}\mathfrak{g} \leq \binom{k}{2}$. This proves (ii). To prove (iii), assume for a contradiction that $\dim Z(\mathfrak{g}) = \dim \mathfrak{g} - 1$. Then $k = 1$, so (ii) forces $\mathcal{D}\mathfrak{g} = \{0\}$. But then $\mathfrak{g}$ is abelian, so $Z(\mathfrak{g}) = \mathfrak{g}$, contrary to the assumption.

Solution to Exercise 3.4.4. Since $[\cdot, \cdot]$ is linear in the first variable, $Z_\mathfrak{g}(S)$ is a subspace. If $x, y \in Z_\mathfrak{g}(S)$, $z \in S$ then

$$[[x, y], z] = [[x, z], y] + [x, [y, z]] = [0, y] + [x, 0] = 0$$

so that $[x, y] \in Z_\mathfrak{g}(S)$, which proves that $Z_\mathfrak{g}(S)$ is a subalgebra as claimed in (i). For (ii), suppose that $\mathfrak{h}$ is an ideal of $\mathfrak{g}$. If $x \in \mathfrak{g}$, $y \in Z_\mathfrak{g}(\mathfrak{h})$, and $z \in \mathfrak{h}$ then

$$[[x, y], z] = [[x, z], y] + [x, [y, z]] = 0 + [x, 0] = 0,$$

so that $[x, y] \in Z_\mathfrak{g}(\mathfrak{h})$; this proves that $Z_\mathfrak{g}(\mathfrak{h})$ is an ideal of $\mathfrak{g}$.

For part (iii) note that, since $\mathfrak{d}_n$ is abelian, $\mathfrak{d}_n \subseteq Z_{\mathfrak{gl}_n}(\mathfrak{d}_n)$. In fact, equality holds here: the quickest way to see this is to notice that if a matrix commutes with all diagonal matrices then it commutes in particular with the diagonal matrix with diagonal entries $1, 2, \ldots, n$; hence it preserves the eigenspaces of this matrix, hence it is diagonal. Otherwise, one can argue more directly: if $x = \sum_{i,j} x_{ij} e_{ij}$ commutes with $\mathfrak{d}_n$ then we have $[e_{ii}, x] = 0$ for all i, which means that $\sum_j x_{ij} e_{ij} - \sum_j x_{ji} e_{ji} = 0$ for all i. Clearly this forces $x_{ij} = 0$ for all $i \neq j$, so x is diagonal.

The proof suggested in (iv) runs as follows. Any element of the centre $Z(\mathfrak{gl}_n)$ must in particular commute with all diagonal matrices, so by (iii) it must be diagonal. If we write it as $\sum_k a_k e_{kk}$ for some $a_1, \ldots, a_n \in \mathbb{C}$, we can then simply observe that $0 = [\sum_k a_k e_{kk}, e_{ij}] = (a_i - a_j) e_{ij}$ for all $i \neq j$, forcing all the diagonal entries a_k to be equal as required.

Solution to Exercise 3.4.5. It is clear that $N_\mathfrak{g}(\mathfrak{h})$ is a subspace of $\mathfrak{g}$. If $x, y \in N_\mathfrak{g}(\mathfrak{h})$ and $z \in \mathfrak{h}$ then

$$[[x, y], z] = [[x, z], y] + [x, [y, z]] \in [\mathfrak{h}, y] + [x, \mathfrak{h}] \subseteq \mathfrak{h} + \mathfrak{h} = \mathfrak{h},$$

so that $[x, y] \in N_\mathfrak{g}(\mathfrak{h})$; this shows that $N_\mathfrak{g}(\mathfrak{h})$ is a subalgebra of $\mathfrak{g}$. Since $\mathfrak{h}$ is a subalgebra, $\mathfrak{h}$ is contained in $N_\mathfrak{g}(\mathfrak{h})$ and is obviously an ideal of $N_\mathfrak{g}(\mathfrak{h})$. Indeed, $N_\mathfrak{g}(\mathfrak{h})$ is by definition the largest subalgebra of $\mathfrak{g}$ in which $\mathfrak{h}$ is an ideal. This proves (i).

Obviously $Z_\mathfrak{g}(\mathfrak{h})$ is a subset of $N_\mathfrak{g}(\mathfrak{h})$. Since $Z_\mathfrak{g}(\mathfrak{h})$ is a subalgebra of $\mathfrak{g}$ by Exercise 3.4.4(i), it is also a subalgebra of $N_\mathfrak{g}(\mathfrak{h})$. If $x \in N_\mathfrak{g}(\mathfrak{h})$, $y \in Z_\mathfrak{g}(\mathfrak{h})$, and $z \in \mathfrak{h}$ then

$$[[x, y], z] = [[x, z], y] + [x, [y, z]] = 0 + [x, 0] = 0,$$

so $[x, y] \in Z_\mathfrak{g}(\mathfrak{h})$, proving (ii).

For (iii), let $x = \sum_{i,j} x_{ij}e_{ij} \in \mathfrak{gl}_n$. We have $x \in N_{\mathfrak{gl}_n}(\mathfrak{n}_n)$ if and only if, for all $k < l$, $[x, e_{kl}] \in \mathfrak{n}_n$. But

$$[x, e_{kl}] = \sum_i x_{ik}e_{il} - \sum_j x_{lj}e_{kj},$$

so $[x, e_{kl}] \in \mathfrak{n}_n$ means that $x_{ik} = 0$ for all $i \geq l$ and $x_{lj} = 0$ for all $j \leq k$. Having this for all $k < l$ means exactly that $x_{ij} = 0$ for all $i > j$, i.e. $x \in \mathfrak{b}_n$. Hence $N_{\mathfrak{gl}_n}(\mathfrak{n}_n) \subseteq \mathfrak{b}_n$. We have already seen that $\mathfrak{n}_n$ is an ideal of $\mathfrak{b}_n$, so $N_{\mathfrak{gl}_n}(\mathfrak{n}_n) = \mathfrak{b}_n$.

Solution to Exercise 3.4.6. As suggested by the hint, we choose a basis x, y, z of $\mathfrak{g}$ such that $\mathcal{D}\mathfrak{g} = \mathbb{C}z$. We then have $[x, y] = az$, $[x, z] = bz$, $[y, z] = cz$ for some $a, b, c \in \mathbb{C}$ that are not all zero. In (i), we are assuming in addition that $z \in Z(\mathfrak{g})$ or, in other words, that $b = c = 0$, which implies $a \neq 0$. After scaling x by a^{-1}, our basis satisfies

$$[x, y] = z, \quad [x, z] = 0, \quad [y, z] = 0.$$

These are exactly the Lie bracket relations satisfied by the basis e_{12}, e_{23}, e_{13} of $\mathfrak{n}_3$, so in this case $\mathfrak{g} \cong \mathfrak{n}_3$.

In part (ii), we are assuming that $z \notin Z(\mathfrak{g})$, so b and c are not both zero. Interchanging x and y if necessary, we can assume that $b \neq 0$. To find the desired isomorphism we need to work out what the centre is: we have $\alpha x + \beta y + \gamma z \in Z(\mathfrak{g})$ if and only if its Lie bracket with each basis vector is 0, i.e.

$$a\beta + b\gamma = -a\alpha + c\gamma = -b\alpha - c\beta = 0.$$

These equations are equivalent to $\alpha = -cb^{-1}\beta$ and $\gamma = -ab^{-1}\beta$, so that $Z(\mathfrak{g}) = \mathbb{C}(-cb^{-1}x + y - ab^{-1}z)$ is one-dimensional. Note that $\mathbb{C}\{x, z\}$ is an ideal of $\mathfrak{g}$ that is complementary to $Z(\mathfrak{g})$, so $\mathfrak{g} \cong Z(\mathfrak{g}) \times \mathbb{C}\{x, z\}$. Since $Z(\mathfrak{g}) \cong \mathbb{C}$ and $\mathbb{C}\{x, z\} \cong \mathfrak{l}_2$, this gives $\mathfrak{g} \cong \mathbb{C} \times \mathfrak{l}_2$ as required.

Solution to Exercise 3.4.7. Choose a basis x, y, z of $\mathfrak{g}$ such that $\mathcal{D}\mathfrak{g} = \mathbb{C}\{y, z\}$. We then have

$$[x, y] = ay + cz,\ [x, z] = by + dz,\ [y, z] = ey + fz \quad \text{for some } a, b, c, d, e, f \in \mathbb{C}.$$

The Jacobi identity tells us that

$$\begin{aligned} 0 &= [[x, y], z] + [[y, z], x] + [[z, x], y] \\ &= [ay + cz, z] + [ey + fz, x] - [by + dz, y] \\ &= a(ey + fz) - e(ay + cz) - f(by + dz) + d(ey + fz) \\ &= (de - bf)y + (af - ce)z, \end{aligned}$$

so $bf = de$ and $af = ce$. These equations mean that $by + dz$ and $ey + fz$ are linearly dependent and $ay + cz$ and $ey + fz$ are linearly dependent. If $ey + fz \neq 0$ then this means that $\mathcal{D}\mathfrak{g} = \mathbb{C}\{ey + fz\}$, contrary to assumption. So $[y, z] = ey + fz = 0$. In other words, $\mathcal{D}\mathfrak{g}$ is abelian. The Lie bracket relations are now in the same form as in Exercise 2.3.8. It was assumed in that exercise that $ad - bc \neq 0$; this holds because $\mathcal{D}\mathfrak{g} = \mathbb{C}\{y, z\}$. We have thus proved that $\mathfrak{g}$ is in one of the isomorphism classes described in Exercise 2.3.8. By Exercise 3.4.3(ii), we have $\dim \mathfrak{g} - \dim Z(\mathfrak{g}) \geq 3$, so $Z(\mathfrak{g}) = \{0\}$.

Solution to Exercise 3.4.8. Part (i) is easy: the composition of the inclusion $\mathfrak{h}' \hookrightarrow \mathfrak{g}$ and the projection $\mathfrak{g} \twoheadrightarrow \mathfrak{g}/\mathfrak{h}$ is a Lie algebra homomorphism $\mathfrak{h}' \to \mathfrak{g}/\mathfrak{h}$, which is bijective by basic linear algebra. One example for part (ii) is $\mathfrak{g} = \mathfrak{n}_3$ and $\mathfrak{h} = \mathcal{D}\mathfrak{n}_3 = \mathbb{C}e_{13}$. If there were a complementary subalgebra $\mathfrak{h}'$, it would be an ideal of $\mathfrak{n}_3$ because $e_{13} \in Z(\mathfrak{n}_3)$. But then $\mathcal{D}\mathfrak{n}_3$ would equal $\mathcal{D}\mathfrak{h} \oplus \mathcal{D}\mathfrak{h}' = \mathcal{D}\mathfrak{h}'$, contradicting the fact that it equals $\mathfrak{h}$.

We now prove (iii). If such a Lie algebra homomorphism $\varphi : \mathfrak{g}/\mathfrak{h} \to \mathfrak{g}$ exists then let $\mathfrak{h}' = \operatorname{im}(\varphi)$. As the image of a homomorphism, $\mathfrak{h}'$ is certainly a subalgebra of $\mathfrak{g}$ so we need only show that it is complementary to $\mathfrak{h}$. The constraint on φ implies that it is injective, so $\dim \mathfrak{h}' = \dim(\mathfrak{g}/\mathfrak{h}) = \dim \mathfrak{g} - \dim \mathfrak{h}$, which means that it is enough to show that $\mathfrak{h} \cap \mathfrak{h}' = \{0\}$. But $\varphi(x + \mathfrak{h}) \in \mathfrak{h}$ can only occur when $x + \mathfrak{h} = 0 + \mathfrak{h}$, in which case $\varphi(x + \mathfrak{h}) = 0$, so that $\mathfrak{h} \cap \mathfrak{h}' = \{0\}$ is true. Conversely, suppose that a complementary subalgebra $\mathfrak{h}'$ exists. Then every element of $\mathfrak{g}/\mathfrak{h}$ can be written uniquely as $x + \mathfrak{h}$ where $x \in \mathfrak{h}'$, and we have a well-defined linear map $\varphi : \mathfrak{g}/\mathfrak{h} \to \mathfrak{g}$ such that $\varphi(x + \mathfrak{h}) = x$ for all $x \in \mathfrak{h}'$. For any $x, x' \in \mathfrak{h}'$, we have $[x + \mathfrak{h}, x' + \mathfrak{h}] = [x, x'] + \mathfrak{h}$ where $[x, x'] \in \mathfrak{h}'$, so $\varphi([x + \mathfrak{h}, x' + \mathfrak{h}]) = [x, x'] = [\varphi(x + \mathfrak{h}), \varphi(x' + \mathfrak{h})]$. Thus φ is a Lie algebra homomorphism as required.

Solution to Exercise 3.4.9. Let x_1 be a nonzero element of $\mathcal{D}\mathfrak{g}$, so that $\mathcal{D}\mathfrak{g} = \mathbb{C}x_1$. In the notation of Exercise 3.4.4, we have

$$Z_{\mathfrak{g}}(\mathcal{D}\mathfrak{g}) = \{x \in \mathfrak{g} \mid [x, x_1] = 0\}.$$

By assumption $x_1 \notin Z(\mathfrak{g})$, so the linear map $\mathfrak{g} \to \mathcal{D}\mathfrak{g} : x \mapsto [x, x_1]$ is nonzero and therefore its image is the whole of $\mathcal{D}\mathfrak{g}$ and its kernel $Z_{\mathfrak{g}}(\mathcal{D}\mathfrak{g})$ has dimension $m - 1$. Note that $\mathcal{D}\mathfrak{g} \subset Z_{\mathfrak{g}}(\mathcal{D}\mathfrak{g})$. Take any $x \in \mathfrak{g}$ and any $y, z \in Z_{\mathfrak{g}}(\mathcal{D}\mathfrak{g})$. The Jacobi identity tells us that

$$[x, [y, z]] = [[x, y], z] + [y, [x, z]] = 0 + 0 = 0.$$

So $[y, z] \in Z(\mathfrak{g})$ and also $[y, z] \in \mathcal{D}\mathfrak{g}$, forcing $[y, z] = 0$. Thus $Z_{\mathfrak{g}}(\mathcal{D}\mathfrak{g})$ is abelian. We can choose $x_2, \ldots, x_m \in \mathfrak{g}$ so that $x_1, \ldots, x_m$ is a basis of $\mathfrak{g}$ and $Z_{\mathfrak{g}}(\mathcal{D}\mathfrak{g})$ is spanned by $x_1, \ldots, x_{m-1}$. Then $[x_i, x_j] = 0$ whenever $i, j \leq m - 1$. For $i \leq m - 1$, we have $[x_i, x_m] = a_i x_1$ for some $a_i \in \mathbb{C}$. Since $x_m \notin Z_{\mathfrak{g}}(\mathcal{D}\mathfrak{g})$, we have $a_1 \neq 0$.

Define a new basis $x'_1, \dots, x'_m$ of $\mathfrak{g}$ by $x'_i = x_i - (a_i/a_1)x_1$ for $2 \leq i \leq m-1$ and $x'_i = x_i$ for $i = 1$ or $i = m$. We still have $[x'_i, x'_j] = 0$ whenever $i, j \leq m-1$, but now $[x'_i, x'_m] = 0$ for $2 \leq i \leq m-1$. So $\mathfrak{g}$ is the direct sum of the ideals $\mathbb{C}\{x'_2, \dots, x'_{m-1}\}$ and $\mathbb{C}\{x'_1, x'_m\}$, the first of which is abelian and the second of which is non-abelian (because $[x'_1, x'_m] = a_1 x'_1 \neq 0$).

Solutions for Chapter 4 exercises

Solution to Exercise 4.5.1. By definition of an $\mathfrak{l}_2$-action, we know that

$$\begin{aligned}
\begin{pmatrix} 0 & 1 & 0 \\ 0 & 0 & 1 \\ 0 & 0 & 0 \end{pmatrix} &= \left[\begin{pmatrix} 1 & 0 & 0 \\ a & b & 0 \\ c & d & e \end{pmatrix}, \begin{pmatrix} 0 & 1 & 0 \\ 0 & 0 & 1 \\ 0 & 0 & 0 \end{pmatrix} \right] \\
&= \begin{pmatrix} 0 & 1 & 0 \\ 0 & a & b \\ 0 & c & d \end{pmatrix} - \begin{pmatrix} a & b & 0 \\ c & d & e \\ 0 & 0 & 0 \end{pmatrix} \\
&= \begin{pmatrix} -a & 1-b & 0 \\ -c & a-d & b-e \\ 0 & c & d \end{pmatrix}.
\end{aligned}$$

Hence $a = b = c = d = 0$ and $e = -1$, meaning that $s_{\mathbb{C}^3}$ has a diagonal matrix with diagonal entries $1, 0, -1$. This answers (i).

To answer (ii), suppose that $v = \alpha v_1 + \beta v_2 + \gamma v_3$. The $\mathfrak{l}_2$-submodule W of $\mathbb{C}^3$ generated by v must also contain $sv = \alpha v_1 - \gamma v_3$ and $s^2 v = \alpha v_1 + \gamma v_3$. It is easy to see that the span $\mathbb{C}\{v, sv, s^2 v\}$ equals the span of those standard basis vectors v_i whose coefficient in v is nonzero. Because W is stable under the action of t, it must contain v_1 if it contains v_2 and it must contain v_2 if it contains v_3. Thus the possibilities are: $W = \{0\}$, $W = \mathbb{C}v_1$, $W = \mathbb{C}\{v_1, v_2\}$, $W = \mathbb{C}^3$.

Solution to Exercise 4.5.2. We found the values of ψ on the standard basis elements of $\mathfrak{sl}_2$ in the solution to Exercise 2.3.5(iii). We are now regarding these as representing matrices for an $\mathfrak{sl}_2$-action on $\mathbb{C}^3$, relative to the standard basis of $\mathbb{C}^3$. The representing matrices for the adjoint representation of $\mathfrak{sl}_2$ on $\mathfrak{sl}_2$, relative to the standard basis, were found in Example 4.1.14. Notice that the two collections of representing matrices are the same except for some signs; we can make them coincide exactly by using a slightly different basis for $\mathbb{C}^3$, namely $-v_1, v_2, v_3$. So the vector space isomorphism $\mathfrak{sl}_2 \xrightarrow{\sim} \mathbb{C}^3$ that sends e to $-v_1$, h to v_2, and f to v_3 is an $\mathfrak{sl}_2$-module isomorphism.

Solution to Exercise 4.5.3. For part (i), if we order the basis of $\mathfrak{gl}_2$ as $e_{11}, e_{12}, e_{21}, e_{22}$ then the representing matrices are

$$e_{11}: \begin{pmatrix} 0 & 0 & 0 & 0 \\ 0 & 1 & 0 & 0 \\ 0 & 0 & -1 & 0 \\ 0 & 0 & 0 & 0 \end{pmatrix}, \quad e_{12}: \begin{pmatrix} 0 & 0 & 1 & 0 \\ -1 & 0 & 0 & 1 \\ 0 & 0 & 0 & 0 \\ 0 & 0 & -1 & 0 \end{pmatrix}, \quad e_{22}: \begin{pmatrix} 0 & 0 & 0 & 0 \\ 0 & -1 & 0 & 0 \\ 0 & 0 & 1 & 0 \\ 0 & 0 & 0 & 0 \end{pmatrix}.$$

For example, the second matrix expresses the following commutators:

$$[e_{12}, e_{11}] = -e_{12}, \quad [e_{12}, e_{12}] = 0, \quad [e_{12}, e_{21}] = e_{11} - e_{22}, \quad [e_{12}, e_{22}] = e_{12}.$$

To answer (ii) we must find a nontrivial $\mathfrak{b}_2$-submodule of $\mathfrak{gl}_2$. The obvious one is $\mathfrak{b}_2$ itself ($\mathfrak{b}_2$ is the span of the first, second, and fourth basis vectors, so in terms of the above matrices this submodule is visible from the fact that the first, second, and fourth entries of the third row are 0).

We will show that the answer to (iii) is negative by proving that there is no $\mathfrak{b}_2$-submodule of $\mathfrak{gl}_2$ that is complementary to $\mathfrak{b}_2$. Since such a submodule would have to be one-dimensional, we just have to find all one-dimensional $\mathfrak{b}_2$-submodules. However, a one-dimensional subspace $\mathbb{C}v$ is a submodule exactly when v is simultaneously an eigenvector for all the representing transformations, so we need to find common eigenvectors of the above matrices. The first matrix has one-dimensional eigenspaces for the eigenvalues 1 and -1 and a two-dimensional eigenspace for the eigenvalue 0. The 1-eigenvector (unique up to a scalar) is also an eigenvector of the other two matrices, and this corresponds to the fact that $\mathfrak{n}_2 = \mathbb{C}e_{12}$ is a one-dimensional $\mathfrak{b}_2$-submodule. The (-1)-eigenvector is not an eigenvector of the second matrix, so we do not get a submodule in this case. The only element of the 0-eigenspace (up to a scalar) that is an eigenvector for the second matrix is the sum of the first and fourth basis elements, and this corresponds to the fact that $Z(\mathfrak{gl}_2) = \mathbb{C}(e_{11} + e_{22})$ is a one-dimensional $\mathfrak{b}_2$-submodule. So the only one-dimensional $\mathfrak{b}_2$-submodules of $\mathfrak{gl}_2$ are $\mathfrak{n}_2$ and $Z(\mathfrak{gl}_2)$ and both are contained in $\mathfrak{b}_2$ rather than complementary to it. So $\mathfrak{gl}_2$ is not a completely reducible $\mathfrak{b}_2$-module.

Solution to Exercise 4.5.4. As seen in Example 4.2.8, classifying two-dimensional $\mathfrak{l}_2$-modules up to isomorphism amounts to classifying pairs of 2×2 matrices (S, T) that satisfy $[S, T] = T$ up to the equivalence relation of simultaneous conjugation, in which two such pairs (S, T) and (S', T') are equivalent if there is some $g \in GL_2$ that satisfies both $gSg^{-1} = S'$ and $gTg^{-1} = T'$. Here S, T are matrices that represent the Lie algebra elements s, t relative to some basis of the $\mathfrak{l}_2$-module.

We can assume without loss of generality that T is in Jordan form since every equivalence class obviously contains pairs with this property. Example 4.2.8 covered the case where this Jordan form is $\left(\begin{smallmatrix} 0 & 1 \\ 0 & 0 \end{smallmatrix}\right)$. Notice that T must have trace zero since it is a commutator, so the only other possibility is that $T = \left(\begin{smallmatrix} \alpha & 0 \\ 0 & -\alpha \end{smallmatrix}\right)$ for $\alpha \in \mathbb{C}$. If we let $S = \left(\begin{smallmatrix} a & b \\ c & d \end{smallmatrix}\right)$, the bracket relation becomes

$$\begin{pmatrix} \alpha & 0 \\ 0 & -\alpha \end{pmatrix} = \left[\begin{pmatrix} a & b \\ c & d \end{pmatrix}, \begin{pmatrix} \alpha & 0 \\ 0 & -\alpha \end{pmatrix} \right] = \begin{pmatrix} 0 & -2b\alpha \\ 2c\alpha & 0 \end{pmatrix},$$

which forces $\alpha = 0$ so that T is the zero matrix. Then any choice of S will satisfy the equation $[S, T] = T$, and our problem boils down to classifying 2×2 matrices up to conjugation, which is solved by the Jordan-form theorem. So representatives of the classes of this type are the pairs $(S, 0)$ where S is in Jordan form. In retrospect, Example 4.2.8 was the interesting part of (i).

In (ii) we are dealing with a two-dimensional abelian Lie algebra $\mathbb{C}\{x, y\}$. As stated in the hint, classifying two-dimensional $\mathbb{C}\{x, y\}$-modules up to isomorphism amounts to classifying pairs of 2×2 matrices (X, Y) that satisfy $[X, Y] = 0$, i.e. X commutes with Y, up to the equivalence relation of simultaneous conjugation. Again we can assume without loss of generality that Y is in Jordan form. For each fixed Jordan-form matrix Y, two pairs (X, Y) and (X', Y) are equivalent if and only if there exists some $g \in GL_2$ such that $gXg^{-1} = X'$ and also $gYg^{-1} = Y$, i.e. g commutes with Y. We will now go through the 2×2 Jordan forms in turn.

- $Y = \left(\begin{smallmatrix} a & 0 \\ 0 & a \end{smallmatrix}\right)$ for some $a \in \mathbb{C}$. In this case every $X \in \mathrm{Mat}_2$ commutes with Y, and so the problem reduces to that of classifying 2×2 matrices up to conjugation by an arbitrary 2×2 invertible matrix, to which the answer is again given by Jordan form. So representatives of the classes of this type are the pairs $(X, \left(\begin{smallmatrix} a & 0 \\ 0 & a \end{smallmatrix}\right))$ where X is in Jordan form.
- $Y = \left(\begin{smallmatrix} a & 0 \\ 0 & b \end{smallmatrix}\right)$ for some distinct $a, b \in \mathbb{C}$. In this case $[X, Y] = 0$ if and only if X is diagonal (in other words, X preserves the eigenspaces of Y). The conjugation action of a diagonal matrix on another diagonal matrix is trivial because diagonal matrices commute with each other, so (X, Y) and (X', Y) are equivalent only when $X = X'$. Representatives of the classes of this type are the pairs $(\left(\begin{smallmatrix} c & 0 \\ 0 & d \end{smallmatrix}\right), \left(\begin{smallmatrix} a & 0 \\ 0 & b \end{smallmatrix}\right))$ with $a \neq b$.
- $Y = \left(\begin{smallmatrix} a & 1 \\ 0 & a \end{smallmatrix}\right)$ for some $a \in \mathbb{C}$. In this case $[X, Y] = 0$ if and only if X has the form $\left(\begin{smallmatrix} b & c \\ 0 & b \end{smallmatrix}\right)$ for $b, c \in \mathbb{C}$, as a short calculation shows. Any two matrices of this form commute, since they are linear combinations of 1_2 and e_{12}, which commute, so again the conjugation action is trivial. Representatives of the classes of this type are the pairs $(\left(\begin{smallmatrix} b & c \\ 0 & b \end{smallmatrix}\right), \left(\begin{smallmatrix} a & 1 \\ 0 & a \end{smallmatrix}\right))$.

Solution to Exercise 4.5.5. Part (i) asks us to prove that

$$[z, [x, y]] = [[z, x], y] + [x, [z, y]] \quad \text{for all } x, y, z \in \mathfrak{g},$$

which is the Jacobi identity. For (ii), note that $\mathrm{ad}(\mathfrak{g})$ is certainly a subalgebra of $\mathrm{Der}(\mathfrak{g})$, since $\mathrm{ad}(\mathfrak{g})$ is the image of the Lie algebra homomorphism ad. If $\varphi \in \mathrm{Der}(\mathfrak{g})$ and $x, y \in \mathfrak{g}$ then

$$[\varphi, \mathrm{ad}(x)](y) = \varphi(\mathrm{ad}(x)(y)) - \mathrm{ad}(x)(\varphi(y)) = \varphi([x, y]) - [x, \varphi(y)] = [\varphi(x), y],$$

so that $[\varphi, \mathrm{ad}(x)] = \mathrm{ad}(\varphi(x)) \in \mathrm{ad}(\mathfrak{g})$. Thus $\mathrm{ad}(\mathfrak{g})$ is an ideal of $\mathrm{Der}(\mathfrak{g})$.

For part (iii), we have already seen in Example 3.1.6 that any element of $\mathrm{Der}(\mathfrak{l}_2)$ is of the form $s \mapsto ct, t \mapsto dt$ for some $c, d \in \mathbb{C}$. Since

$$[ds - ct, s] = ct, \quad [ds - ct, t] = dt,$$

such a derivation is exactly of the form $\mathrm{ad}(ds - ct)$.

An example of a Lie algebra as in (iv) is $\mathfrak{g} = \mathbb{C} \times \mathfrak{l}_2$. We will prove that the linear transformation φ of $\mathfrak{g}$ given by $\varphi(a, bs + ct) = (a, 0)$ is in $\mathrm{Der}(\mathfrak{g})$ but not $\mathrm{ad}(\mathfrak{g})$. To see that φ is a derivation, we calculate as follows:

$$\varphi([(a, bs + ct), (a', b's + c't)]) = \varphi(0, (bc' - cb')t) = (0, 0),$$

and

$$[\varphi(a, bs + ct), (a', b's + c't)] = (0, 0) = [(a, bs + ct), \varphi(a', b's + c't)].$$

Now we have $\mathrm{ad}(a, bs+ct)(a', b's+c't) = (0, (bc' - cb')t)$, so φ is not of the form $\mathrm{ad}(a, bs + ct)$ for any a, b, c.

Solution to Exercise 4.5.6. This bracket is clearly bilinear and skew-symmetric, so we just need to check the Jacobi identity. For $x, x', x'' \in \mathfrak{g}$ and $v, v', v'' \in V$, we have

$$\begin{aligned}
[[(x, v), (x', v')], (x'', v'')] &= [([x, x'], xv' - x'v), (x'', v'')] \\
&= ([[x, x'], x''], [x, x']v'' - x''(xv' - x'v)) \\
&= ([[x, x'], x''], xx'v'' - x'xv'' - x''xv' + x''x'v).
\end{aligned}$$

If we cyclically permute $(x, v), (x', v'), (x'', v'')$ and add the three resulting expressions, the first component vanishes by the Jacobi identity in $\mathfrak{g}$ and the second component vanishes by simple cancellation. This proves (i).

An example of a Lie algebra as in (ii) is $\mathfrak{h} = \mathfrak{sl}_2 \ltimes \mathbb{C}^2$, where we regard $\mathbb{C}^2$ as an $\mathfrak{sl}_2$-module via the natural representation. For notational convenience we will identify $(x, 0)$ with x for any $x \in \mathfrak{sl}_2$ and $(0, v)$ with v for any $v \in \mathbb{C}^2$. Then $\mathfrak{h}$ has basis e, h, f, v_1, v_2. From the definition in (i) we see that the bracket relations among e, h, f are the same as the familiar ones in $\mathfrak{sl}_2$, that $[v_1, v_2] = 0$, and that $[x, v] = xv$ for any $x \in \mathfrak{sl}_2$ and $v \in \mathbb{C}^2$. Explicitly,

$$[e, v_1] = 0,\ [e, v_2] = v_1,\ [h, v_1] = v_1,\ [h, v_2] = -v_2,\ [f, v_1] = v_2,\ [f, v_2] = 0.$$

So $\mathcal{D}\mathfrak{h}$ contains e, h, f, v_1, v_2 and hence $\mathcal{D}\mathfrak{h} = \mathfrak{h}$. Suppose that $ae+bh+cf+a_1v_1+a_2v_2 \in Z(\mathfrak{h})$. Then

$$0 = [h, ae + bh + cf + a_1v_1 + a_2v_2] = 2ae - 2cf + a_1v_1 - a_2v_2$$

and

$$0 = [e, ae + bh + cf + a_1v_1 + a_2v_2] = -2be + ch + a_2v_1,$$

forcing $a = b = c = a_1 = a_2 = 0$. So $Z(\mathfrak{h}) = \{0\}$. Clearly $\mathbb{C}\{v_1, v_2\}$ is an ideal of $\mathfrak{h}$, so $\mathfrak{h}$ is not simple. Finally, $\mathfrak{h}$ cannot be a direct product of more than one simple Lie algebra since $\dim \mathfrak{h} = 5$ and any simple Lie algebra has dimension ≥ 3.

Solution to Exercise 4.5.7. If $\mathfrak{g}$ had a nontrivial ideal $\mathfrak{h}$ then we would be in the situation of Exercise 3.4.3(i), contrary to the assumption that $\mathcal{D}\mathfrak{g} = \mathfrak{g}$. So (i) is proved. We can prove (ii) by contradiction, assuming that every $\mathrm{ad}_\mathfrak{g}(x)$ has zero as its only eigenvalue. Choose any nonzero $x \in \mathfrak{g}$. Notice that the 0-eigenspace (i.e. kernel) of $\mathrm{ad}_\mathfrak{g}(x)$ contains x itself. It cannot be the case that x belongs to the image of $\mathrm{ad}_\mathfrak{g}(x)$, because the equation $[x, y] = x$ would say that x is a (-1)-eigenvector of $\mathrm{ad}_\mathfrak{g}(y)$. So the Jordan form of $\mathrm{ad}_\mathfrak{g}(x)$ is not just a single Jordan block, and its 0-eigenspace must have dimension greater than 1. Hence there is an element $y \in \mathfrak{g}$ such that $[x, y] = 0$ and x, y are linearly independent. But then we could extend x, y to a basis x, y, z of $\mathfrak{g}$ and $\mathcal{D}\mathfrak{g}$ would be spanned by the two elements $[x, z]$ and $[y, z]$, contrary to the assumption that $\mathcal{D}\mathfrak{g} = \mathfrak{g}$. This proves (ii).

Now let $x \in \mathfrak{g}$ be such that $\mathrm{ad}_\mathfrak{g}(x)$ has a nonzero eigenvalue. Proposition 4.1.8 applied to the adjoint representation of $\mathfrak{g}$ shows that $\mathrm{tr}(\mathrm{ad}_\mathfrak{g}(x)) = 0$, so the sum of the eigenvalues of $\mathrm{ad}_\mathfrak{g}(x)$ is 0. Moreover, 0 is an eigenvalue of $\mathrm{ad}_\mathfrak{g}(x)$, since x belongs to its 0-eigenspace. This forces the eigenvalues of $\mathrm{ad}_\mathfrak{g}(x)$ to be d, 0, and $-d$ for some nonzero d. If we set $h = 2d^{-1}x$ then $\mathrm{ad}_\mathfrak{g}(h)$ has eigenvalues 2, 0, and -2; letting e be a 2-eigenvector of $\mathrm{ad}_\mathfrak{g}(h)$ and f a (-2)-eigenvector, we know that e, h, f is a basis of $\mathfrak{g}$ satisfying $[h, e] = 2e$, $[h, f] = -2f$. Now suppose $[e, f] = ae + bh + cf$, where $a, b, c \in \mathbb{C}$. The Jacobi identity tells us that

$$0 = [h, [e, f]] - [[h, e], f] - [e, [h, f]] = 2ae - 2cf - 2[e, f] + 2[e, f] = 2ae - 2cf,$$

so $a = c = 0$. If b were also 0 then $\mathcal{D}\mathfrak{g}$ would not include h, contrary to assumption, so $b \neq 0$. Replacing f by $b^{-1}f$ we get a basis satisfying exactly the $\mathfrak{sl}_2$ relations, so $\mathfrak{g} \cong \mathfrak{sl}_2$.

Solution to Exercise 4.5.8. We prove (i) by induction on i. The base case, when $i = 0$, says that for all $y \in \mathfrak{h}$ we have $yv - \chi(y)v = 0$, which is true by the definition of $V_\chi^\mathfrak{h}$. Suppose that $i \geq 1$ and that we know the result for $i - 1$. For all $y \in \mathfrak{h}$, this induction hypothesis tells us that

$$yx^{i-1}v - \chi(y)x^{i-1}v \in \mathbb{C}\{v, xv, \ldots, x^{i-2}v\}$$

and also

$$[y, x]x^{i-1}v - \chi([y, x])x^{i-1}v \in \mathbb{C}\{v, xv, \ldots, x^{i-2}v\}.$$

These inclusions imply, respectively, that

$$xyx^{i-1}v - \chi(y)x^i v \in \mathbb{C}\{xv, x^2v, \ldots, x^{i-1}v\}$$

and

$$yx^i v - xyx^{i-1}v \in \mathbb{C}\{v, xv, \dots, x^{i-1}v\}.$$

Adding the latter two inclusions gives $yx^i v - \chi(y)x^i v \in \mathbb{C}\{v, xv, \dots, x^{i-1}v\}$, completing the induction step.

To prove (ii), let $y \in \mathfrak{h}$, $x \in \mathfrak{g}$ and choose any nonzero $v \in V_\chi^{\mathfrak{h}}$. As we form the sequence $v, xv, x^2v, \dots$ there must come a time when $x^m v$ is linearly dependent on the earlier members of the sequence; let m be the minimal such exponent. Then $v, xv, x^2v, \dots, x^{m-1}v$ form a basis of the subspace W that they span, and x_V preserves the subspace W. The result of (i) implies that y_V also preserves the subspace W, and the matrix of y_W relative to the basis $v, xv, x^2v, \dots, x^{m-1}v$ is upper-triangular with diagonal entries all equal to $\chi(y)$. But $[y, x]$ also belongs to $\mathfrak{h}$, so we also know that the matrix of $[y, x]_W$ relative to this basis is upper-triangular with diagonal entries all equal to $\chi([y, x])$. Since $[y, x]_W = [y_W, x_W]$, it has trace zero. So $m\chi([y, x]) = 0$, giving $\chi([y, x]) = 0$ as required.

We can now prove (iii). Let $x \in \mathfrak{g}$ and $v \in V_\chi^{\mathfrak{h}}$. We must show that $xv \in V_\chi^{\mathfrak{h}}$, which means that $yxv = \chi(y)xv$ for all $y \in \mathfrak{h}$. This follows from (ii):

$$yxv = [y, x]v + xyv = \chi([y, x])v + x\chi(y)v = 0v + \chi(y)xv = \chi(y)xv.$$

Solutions for Chapter 5 exercises

Solution to Exercise 5.3.1. By Corollary 5.2.3(3), when V is written as a direct sum of submodules isomorphic to $V(m)$ then the number of times that a given highest weight m occurs is $\dim V_m - \dim V_{m+2}$. In the present case, this means that $V(5)$ occurs once, $V(4)$ occurs once, $V(3)$ occurs once, $V(2)$ and $V(1)$ do not occur, and $V(0)$ occurs twice. Hence

$$V \cong V(5) \oplus V(4) \oplus V(3) \oplus V(0) \oplus V(0).$$

Solution to Exercise 5.3.2. In our current terminology, Exercise 2.3.5(iii) asks for an explicit isomorphism between two three-dimensional $\mathfrak{sl}_2$-modules. One of these $\mathfrak{sl}_2$-modules is $\mathbb{C}^3$ with $x \in \mathfrak{sl}_2$ represented by the matrix $\psi(x)$ where $\psi : \mathfrak{gl}_2 \to \mathfrak{gl}_3$ is the homomorphism from Example 1.3.1. The other $\mathfrak{sl}_2$-module is $\mathbb{C}^3$ with $x \in \mathfrak{sl}_2$, represented by the matrix $\varphi(x)$ where $\varphi : \mathfrak{sl}_2 \xrightarrow{\sim} \mathfrak{so}_3$ is the isomorphism found in Exercise 2.3.5(ii). Call these modules V and W respectively.

From the results of this chapter, we know that the key to finding a module isomorphism is to consider the weights and weight vectors of the two modules. Since $\psi(h)$ is diagonal, we can see immediately that the weights of V are $2, 0, -2$ and the corresponding weight vectors are the standard basis vectors. Our previous calculation of the eigenvalues and eigenvectors of $\varphi(h)$ was really about establishing that the weights of W are also $2, 0, -2$ and finding the corresponding weight vectors. From what we now know, the mere fact that the weights of V are $2, 0, -2$ implies that we have an $\mathfrak{sl}_2$-module isomorphism $V \cong V(2)$ and similarly that $W \cong V(2)$; thus $V \cong W$. Moreover, to make the isomorphism explicit we just have to find a string basis of V and of W. In V, a highest-weight vector is v_1 and the string it generates is v_1, v_2, v_3. In W, from our previous calculation of the 2-eigenvector of $\varphi(h)$ we see that a highest-weight vector is $\mathbf{i}v_1 + v_2$. Applying $\varphi(f)$ we generate the string $\mathbf{i}v_1 + v_2, 2v_3, \mathbf{i}v_1 - v_2$. (Note how this calculation automatically finds the other eigenvectors of $\varphi(h)$.) So the linear map $V \to W$ sending v_1 to $\mathbf{i}v_1 + v_2$, v_2 to $2v_3$, and v_3 to $\mathbf{i}v_1 - v_2$ is an $\mathfrak{sl}_2$-module isomorphism. The matrix of this linear map is the same $g \in GL_3$ that we found in Exercise 2.3.5(iii).

Solution to Exercise 5.3.3. Let $\pi : \mathfrak{sl}_2 \to \mathfrak{gl}_n$ be the representation of $\mathfrak{sl}_2$ on $\mathbb{C}^n$ described in the question. Then $\pi(e) = e_{12} + e_{23} + \cdots + e_{n-1,n}$, and $\pi(h) = a_1 e_{11} + a_2 e_{22} + \cdots + a_n e_{nn}$ for some $a_i \in \mathbb{C}$. We have

$$\begin{aligned} 2\pi(e) = [\pi(h), \pi(e)] &= \sum_{i=1}^{n} a_i [e_{ii}, e_{12} + e_{23} + \cdots + e_{n-1,n}] \\ &= (a_1 - a_2)e_{12} + (a_2 - a_3)e_{23} + \cdots + (a_{n-1} - a_n)e_{n-1,n}, \end{aligned}$$

which forces $a_1 - a_2 = a_2 - a_3 = \cdots = a_{n-1} - a_n = 2$. Moreover, $\operatorname{tr}(\pi(h)) = \operatorname{tr}([\pi(e), \pi(f)]) = 0$, so the arithmetic progression $a_1, a_2, \ldots, a_n$ must be centred on zero. This tells us that $a_i = n + 1 - 2i$ for all i. Hence the weights of the $\mathfrak{sl}_2$-module $\mathbb{C}^n$ are $n-1, n-3, \ldots, -(n-1)$, which implies that $\mathbb{C}^n \cong V(n-1)$. The weight vectors of $\mathbb{C}^n$ are the standard basis vectors (up to a scalar), so there is a string basis of $\mathbb{C}^n$ of the form $b_1 v_1, b_2 v_2, \ldots, b_n v_n$ for some $b_i \in \mathbb{C}^\times$. Comparing $\pi(e)$ with the formula for the action of e on a string we see that, for $1 \le i \le n-1$,

$$(n - i) b_i v_i = \pi(e) b_{i+1} v_{i+1} = b_{i+1} v_i.$$

So $b_{i+1}/b_i = n - i$. We may as well set $b_1 = 1$, and then the full string is $v_1, (n-1)v_2, (n-1)(n-2)v_3, \ldots, (n-1)!v_n$. Recalling the action of f on a string, we obtain

$$\pi(f)v_i = \begin{cases} i(n-i)v_{i+1} & \text{if } 1 \le i \le n-1, \\ 0 & \text{if } i = n. \end{cases}$$

So $\pi(f) = \sum_{i=1}^{n-1} i(n-1)e_{i+1,i}$ is the matrix whose entries everywhere except on the sub-diagonal are zero, and whose sub-diagonal entries are $1(n-1), 2(n-2), \ldots, (n-2)2, (n-1)1$.

Solution to Exercise 5.3.4. To show (i), we just have to check the Lie bracket relations. The relation $[h, e] = 2e$ corresponds to the assertion that

$$t\frac{\partial}{\partial t}\left(t\frac{\partial p}{\partial u}\right) - u\frac{\partial}{\partial u}\left(t\frac{\partial p}{\partial u}\right) - t\frac{\partial}{\partial u}\left(t\frac{\partial p}{\partial t} - u\frac{\partial p}{\partial u}\right) = 2t\frac{\partial p}{\partial u}.$$

Using the product rule, the left-hand side becomes

$$t\frac{\partial p}{\partial u} + t^2\frac{\partial^2 p}{\partial t\,\partial u} - tu\frac{\partial^2 p}{\partial u^2} - t^2\frac{\partial^2 p}{\partial u\,\partial t} + t\frac{\partial p}{\partial u} + tu\frac{\partial^2 p}{\partial u^2},$$

which simplifies to the right-hand side. The same calculation with t and u switched shows the relation required by $[h, f] = -2f$. Finally, the relation $[e, f] = h$ corresponds to the assertion that

$$t\frac{\partial}{\partial u}\left(u\frac{\partial p}{\partial t}\right) - u\frac{\partial}{\partial t}\left(t\frac{\partial p}{\partial u}\right) = t\frac{\partial p}{\partial t} - u\frac{\partial p}{\partial u},$$

which is also easily checked. This proves (i).

It is immediate from the definitions that $ht^m = mt^m$ and $et^m = 0$, so that t^m is a highest-weight vector of weight m. If we let $w_0, w_1, \ldots, w_m$ denote the string it generates then we can prove by induction on i that $w_i = \binom{m}{i} t^{m-i} u^i$. The $i = 0$ base case is true by definition and, if $w_{i-1} = \binom{m}{i-1} t^{m-i+1} u^{i-1}$, then

$$\begin{aligned} w_i &= \frac{1}{i} f w_{i-1} \\ &= \frac{1}{i}\binom{m}{i-1} u\frac{\partial}{\partial t}(t^{m-i+1}u^{i-1}) \\ &= \frac{m-i+1}{i}\binom{m}{i-1} t^{m-i}u^i \\ &= \binom{m}{i} t^{m-i}u^i, \end{aligned}$$

as required to prove (ii).

Solution to Exercise 5.3.5. To show (i), we just have to check the Lie bracket relations. The relation $[h, e] = 2e$ corresponds to the assertion that, for any $I \subseteq \{1, 2, \ldots, n\}$,

$$\sum_{\substack{J \supset I \\ |J \setminus I| = 1}} (2|J| - n)v_J - (2|I| - n) \sum_{\substack{J \supset I \\ |J \setminus I| = 1}} v_J = 2 \sum_{\substack{J \supset I \\ |J \setminus I| = 1}} v_J,$$

which is clearly true; the relation $[h, f] = -2f$ follows similarly. The relation $[e, f] = h$ corresponds to the assertion that, for any $I \subseteq \{1, 2, \ldots, n\}$,

$$\sum_{\substack{K \subset I \\ |I \setminus K|=1}} \sum_{\substack{J \supset K \\ |J \setminus K|=1}} v_J - \sum_{\substack{J \supset I \\ |J \setminus I|=1}} \sum_{\substack{K \subset J \\ |J \setminus K|=1}} v_K = (2|I| - n)v_I.$$

The coefficient of v_I in the first double sum is $|I|$, since there are $|I|$ choices for the element of I to be removed to form K. The coefficient of v_I in the second double sum is $n-|I|$, since there are $n-|I|$ choices for the element to be added to I to form J. So the coefficient of v_I on the left-hand side is indeed $2|I|-n$. For any other subset L, v_L can occur in the left-hand side only if $|L| = |I| = 1 + |I \cap L|$. In that case, however, v_L will occur in a unique term of the first double sum, corresponding to $K = I \cap L$, and a unique term of the second double sum, corresponding to $J = I \cup L$. So the coefficient of every other v_L in the left-hand side is indeed zero, and this completes the proof.

For (ii) we need to find the weights of V and their multiplicities. Since the basis vector v_I is a weight vector of weight $2|I| - n$, this is easy; the weights of V are $n, n-2, n-4, \ldots, -(n-2), -n$, with respective multiplicities $\binom{n}{n}, \binom{n}{n-1}, \binom{n}{n-2}, \ldots, \binom{n}{1}, \binom{n}{0}$. By Corollary 5.2.3(3), the multiplicity of the irreducible submodule $V(m)$, for $m \geq 0$ of the same parity as n, is

$$\binom{n}{\frac{1}{2}(m+n)} - \binom{n}{\frac{1}{2}(m+n+2)}.$$

Solution to Exercise 5.3.6. Since $s_V t_V - t_V s_V = t_V$, we have

$$(s_V - a - 1)t_V = t_V(s_V - a)$$

for any $a \in \mathbb{C}$. Clearly this implies that $(s_V - a - 1)^k t_V = t_V(s_V - a)^k$ for all $k \in \mathbb{N}$. If $v \in V_a^{\text{gen}}$ then $(s_V - a)^k(v) = 0$ for some $k \in \mathbb{N}$, so we have $(s_V - a - 1)^k(tv) = t0 = 0$, implying that $tv \in V_{a+1}^{\text{gen}}$. This proves (i).

Since s_V has only finitely many eigenvalues, there must be some positive integer N such that, for every eigenvalue a of s_V, $a + N$ is not an eigenvalue of s_V. By (i) we have $t^N v = 0$ for every nonzero generalized eigenvector v of s_V, so $t_V^N = 0$. This proves (ii). To prove (iii), let a be an eigenvalue of s_V such that $a + 1$ is not an eigenvalue of s_V, and let v be an a-eigenvector of s_V. Then $tv = 0$ by (i); so v is also an eigenvector of t_V, meaning that $\mathbb{C}v$ is a one-dimensional $\mathfrak{l}_2$-submodule.

Thus, even though $\mathfrak{l}_2$ is not abelian it has the property shown for abelian Lie algebras in Proposition 4.3.8, that its only irreducible modules are the 1-dimensional ones. Lie algebras with this property are called *solvable*.

Solution to Exercise 5.3.7. We know that we have a direct sum decomposition $V = V_1 \oplus V_2 \oplus \cdots \oplus V_s$, where each V_i is an irreducible $\mathfrak{sl}_2$-submodule that is isomorphic to $V(m_i)$ for some $m_i \in \mathbb{N}$. From the structure of the string bases of these submodules V_i, it is clear that the highest-weight vectors of weight m in V are the nonzero linear combinations of the highest-weight vectors in those V_i such that $m_i = m$. So $V_{[m]}$ is exactly the sum of those V_i such that $m_i = m$ (and is zero if there is no such i). This proves (i).

Note that $V_{[m]}$ has weights $m, m-2, \ldots, -m$ and, for any string $w_0, w_1, \ldots, w_m$ with highest weight m, we have $w_i \in (V_{[m]})_{m-2i}$. Since V is the direct sum of the submodules $V_{[m]}$ and each such submodule is a direct sum of its weight spaces, we can define a linear transformation τ of V by specifying its values on the individual weight spaces:

$$\tau(v) = \begin{cases} \frac{i!}{(m-i)!} f^{m-2i} v & \text{if } v \in (V_{[m]})_{m-2i} \text{ for } i \leq \frac{1}{2}m, \\ \frac{i!}{(m-i)!} e^{2i-m} v & \text{if } v \in (V_{[m]})_{m-2i} \text{ for } i > \frac{1}{2}m. \end{cases}$$

These definitions are specifically chosen, in the light of the formulas for the action of e and f on a string, so as to imply that if $w_0, w_1, \ldots, w_m$ is a string then $\tau(w_i) = w_{m-i}$. The uniqueness of τ follows from the fact that V has a basis consisting of a union of strings. This proves (ii). For the same reason, it suffices to prove (iii) when v is an element of a string, in which case (iii) is clear from the rule $\tau(w_i) = w_{m-i}$ and the symmetry in the string formulas.

Solution to Exercise 5.3.8. We need to ensure that $[h_W, e_W] = 2e_W$, $[h_W, f_W] = -2f_W$, and $[e_W, f_W] = h_W$. The second of these follows from

$$\begin{aligned} [h_W, f_W](w_i) &= h_W(w_{i-1}) - f_W((a+2i)w_i) \\ &= (a + 2i - 2 - a - 2i)w_{i-1} = -2f_W(w_i). \end{aligned}$$

As in Proposition 5.1.7, the condition $[h_W, e_W] = 2e_W$ implies that e_W maps the $(a+2i)$-eigenspace of h_W into the $(a+2i+2)$-eigenspace of h_W, so we are forced to set $e_W(w_i) = b_i w_{i+1}$ for some $b_i \in \mathbb{C}$ and, given this, $[h_W, e_W] = 2e_W$ follows. The condition $[e_W, f_W] = h_W$ is then equivalent to requiring that the following holds for all $i \in \mathbb{Z}$:

$$(a+2i)w_i = e_W(w_{i-1}) - f_W(b_i w_{i+1}) = (b_{i-1} - b_i)w_i$$

or, in other words, $b_i = b_{i-1} - (a+2i)$ for all $i \in \mathbb{Z}$. If we stipulate that $b_0 = b$ as in the question, this system of equations clearly has the unique solution $b_i = b - ia - i(i+1)$. This proves (i).

For (ii), suppose that U is a nonzero $\mathfrak{sl}_2$-submodule of W. Let w be a nonzero element of U, and let $U' \subseteq U$ be the span of all $h^s w$ for $s \geq 1$. Since w is a (finite) linear combination of the vectors w_i, and the latter vectors are eigenvectors for h_W, U' is contained in a span U'' of finitely many w_i. Now U'' is finite-dimensional and $h_{U''}$ is diagonalizable with distinct eigenvalues, so the only eigenvectors for h_W in

U'', up to a scalar, are the vectors w_i that are contained in U''. Since U' is finite-dimensional and h_W-stable it contains some eigenvector for h_W, so it contains w_i for some i as required. Continuing with this train of thought, we see that U must also contain $w_{i'} = f^{i-i'}w_i$, for all $i' < i$, and $e^s w_i$ for all $s \geq 1$. If $b = ja + j(j+1)$ does not hold for any $j \in \mathbb{Z}$ then, by the formula proved above, ew_j is a nonzero multiple of w_{j+1} for all $j \in \mathbb{Z}$. Hence U contains w_{i+s} for $s \geq 1$; so $U = W$, proving that in this case W is irreducible. If, however, $b = ja + j(j+1)$ for some $j \in \mathbb{Z}$ then $ew_j = 0$, from which it is clear that $\mathbb{C}\{w_i \mid i \leq j\}$ is an $\mathfrak{sl}_2$-submodule; so in this case W is reducible. This proves (iii).

Solutions for Chapter 6 exercises

Solution to Exercise 6.5.1. For (i), a general weight vector of weight $m+n-2k$, $0 \leq k \leq n$, in $V(m) \otimes V(n)$ is of the form $\sum_{j=0}^{k} a_j(v_{k-j} \otimes w_j)$ for some $a_j \in \mathbb{C}$. For this to be a highest-weight vector, we must have

$$\begin{aligned} 0 &= e\left(\sum_{j=0}^{k} a_j v_{k-j} \otimes w_j\right) \\ &= \sum_{j=0}^{k-1} a_j(m-k+j+1)v_{k-j-1} \otimes w_j + \sum_{j=1}^{k} a_j(n-j+1)v_{k-j} \otimes w_{j-1} \\ &= \sum_{j=0}^{k-1} \Big(a_j(m-k+j+1) + a_{j+1}(n-j)\Big)(v_{k-j-1} \otimes w_j), \end{aligned}$$

which means exactly that $a_j(m-k+j+1) + a_{j+1}(n-j) = 0$, or in other words

$$a_{j+1} = -\frac{m-k+j+1}{n-j}a_j \quad \text{for all } 0 \leq j \leq k-1.$$

If we set $a_0 = 1$ then

$$\begin{aligned} a_j &= (-1)^j \frac{(m-k+j)(m-k+j-1)\cdots(m-k+1)}{(n-j+1)(n-j+2)\cdots n} \\ &= (-1)^j \frac{(m-k+j)!(n-j)!}{(m-k)!n!}. \end{aligned}$$

So the unique (up to a scalar) highest-weight vector of weight $m+n-2k$ is

$$\sum_{j=0}^{k} (-1)^j \frac{(m-k+j)!(n-j)!}{(m-k)!(n-k)!} v_{k-j} \otimes w_j$$

(here the coefficients a_j have been multiplied by $n!/(n-k)!$ to make them more symmetrical in j and $k-j$).

For (ii), we have

$$\begin{aligned}\frac{1}{i!}f^i(v_0 \otimes w_0) &= \frac{1}{i!}\sum_{j=0}^{i}\binom{i}{j} f^{i-j}(v_0) \otimes f^j(w_0) \\ &= \frac{1}{i!}\sum_{j=0}^{i}\binom{i}{j}(i-j)!v_{i-j} \otimes j!w_j \\ &= \sum_{j=0}^{i} v_{i-j} \otimes w_j.\end{aligned}$$

Here all v_{i-j} for $i-j > m$ and w_j for $j > n$ are to be regarded as 0. So, in terms of the picture in Example 6.1.10, the string basis of this submodule is formed simply by adding the basis vectors on each diagonal.

Solution to Exercise 6.5.2. In Example 6.3.10 we found the representing matrices of the adjoint representation of $\mathfrak{b}_2$. In particular, this showed that $\mathrm{ad}_{\mathfrak{b}_2}(e_{11})$ has trace 1. By Proposition 6.1.2 the matrix representing e_{11} on the dual $\mathfrak{b}_2^*$ of the adjoint representation must have trace -1. So $\mathfrak{b}_2$ and $\mathfrak{b}_2^*$ are not isomorphic as $\mathfrak{b}_2$-modules, which is equivalent to saying that there is no $\mathfrak{b}_2$-invariant nondegenerate bilinear form on $\mathfrak{b}_2$.

Solution to Exercise 6.5.3. Let $\mathfrak{h}$ be an abelian ideal of $\mathfrak{g}$, $x \in \mathfrak{h}$ and $y \in \mathfrak{g}$. By the definition of $\kappa_{\mathfrak{g}}$ (Definition 6.3.9), we need to show that $\mathrm{tr}(\mathrm{ad}_{\mathfrak{g}}(x)\,\mathrm{ad}_{\mathfrak{g}}(y)) = 0$. Choose a basis $x_1, \dots, x_\ell$ of $\mathfrak{h}$, and choose $x_{\ell+1}, \dots, x_m \in \mathfrak{g}$ such that $x_1, \dots, x_m$ is a basis of $\mathfrak{g}$. Consider the matrices of the linear transformations $\mathrm{ad}_{\mathfrak{g}}(x)$ and $\mathrm{ad}_{\mathfrak{g}}(y)$ relative to this basis. Since $x \in \mathfrak{h}$ and $\mathfrak{h}$ is an ideal, $\mathrm{im}(\mathrm{ad}_{\mathfrak{g}}(x)) \subseteq \mathfrak{h}$; since $\mathfrak{h}$ is moreover abelian, $\mathrm{ad}_{\mathfrak{g}}(x)$ acts on $\mathfrak{h}$ as the zero linear transformation. Therefore the matrix of $\mathrm{ad}_{\mathfrak{g}}(x)$ has the block strictly upper-triangular form $\left(\begin{smallmatrix}0 & *\\ 0 & 0\end{smallmatrix}\right)$, where the asterisk indicates an $\ell \times (m-\ell)$ block of unknown entries. Since $\mathfrak{h}$ is an ideal, $\mathrm{ad}_{\mathfrak{g}}(y)$ preserves the subspace $\mathfrak{h}$ so its matrix has the block upper-triangular form $\left(\begin{smallmatrix}* & *\\ 0 & *\end{smallmatrix}\right)$. Therefore $\mathrm{ad}_{\mathfrak{g}}(x)\,\mathrm{ad}_{\mathfrak{g}}(y)$ has the block strictly upper-triangular form $\left(\begin{smallmatrix}0 & *\\ 0 & 0\end{smallmatrix}\right)$, and $\mathrm{tr}(\mathrm{ad}_{\mathfrak{g}}(x)\,\mathrm{ad}_{\mathfrak{g}}(y)) = 0$ as required. This proves the easier direction of Cartan's semisimplicity criterion: if $\kappa_{\mathfrak{g}}$ is nondegenerate then $\mathfrak{g}$ has no nonzero abelian ideals.

Solution to Exercise 6.5.4. We will identify $\mathfrak{g}$ and V with the corresponding subspaces of $\mathfrak{g} \ltimes V$ in the usual way. For (i) note that, by definition of the Lie bracket on $\mathfrak{g} \ltimes V$, V is an abelian ideal of $\mathfrak{g} \ltimes V$. So, by Exercise 6.5.3, $\kappa_{\mathfrak{g}\ltimes V}(x, v) = \kappa_{\mathfrak{g}\ltimes V}(v, v') = 0$ for any $x \in \mathfrak{g}$, $v, v' \in V$. This immediately answers the question about nondegeneracy: $\kappa_{\mathfrak{g}\ltimes V}$ can only be nondegenerate when $V = \{0\}$, in which case $\mathfrak{g} \ltimes V$ is obviously isomorphic to $\mathfrak{g}$ and the nondegeneracy of $\kappa_{\mathfrak{g}\ltimes V}$ is equivalent to that of $\kappa_{\mathfrak{g}}$.

To complete the description of $\kappa_{\mathfrak{g}\ltimes V}$, it suffices to compute $\kappa_{\mathfrak{g}\ltimes V}(x, y)$ for $x, y \in \mathfrak{g}$. If we use a basis of $\mathfrak{g} \ltimes V$ that consists of a basis of $\mathfrak{g}$ followed by a basis of V then, for all $x \in \mathfrak{g}$, $\mathrm{ad}_{\mathfrak{g}\ltimes V}(x)$ has the block diagonal form $\left(\begin{smallmatrix} \mathrm{ad}_{\mathfrak{g}}(x) & 0 \\ 0 & x_V \end{smallmatrix}\right)$, where we are abusing notation by letting $\mathrm{ad}_{\mathfrak{g}}(x)$ and x_V stand for their matrices relative to the chosen bases of $\mathfrak{g}$ and V respectively. So we find that

$$\begin{aligned}\kappa_{\mathfrak{g}\ltimes V}(x, y) &= \mathrm{tr}\left(\left(\begin{smallmatrix} \mathrm{ad}_{\mathfrak{g}}(x) & 0 \\ 0 & x_V \end{smallmatrix}\right)\left(\begin{smallmatrix} \mathrm{ad}_{\mathfrak{g}}(y) & 0 \\ 0 & y_V \end{smallmatrix}\right)\right) \\ &= \mathrm{tr}(\mathrm{ad}_{\mathfrak{g}}(x)\,\mathrm{ad}_{\mathfrak{g}}(y)) + \mathrm{tr}(x_V y_V) = \kappa_{\mathfrak{g}}(x, y) + B_V(x, y).\end{aligned}$$

For part (ii), we will prove the nondegeneracy first. Suppose that $(x, f) \in \mathfrak{g} \ltimes \mathfrak{g}^*$ is such that $B((x, f), (x', f')) = 0$ for all $(x', f') \in \mathfrak{g} \ltimes \mathfrak{g}^*$. Then, taking $f' = 0$, we deduce that $f(x') = 0$ for all $x' \in \mathfrak{g}$, so that $f = 0$; taking $x' = 0$, we deduce that $f'(x) = 0$ for all $f' \in \mathfrak{g}^*$, so that $x = 0$. Hence B is nondegenerate. To show that B is $(\mathfrak{g} \ltimes \mathfrak{g}^*)$-invariant, note that, for $x, x', x'' \in \mathfrak{g}$, $f, f', f'' \in \mathfrak{g}^*$,

$$\begin{aligned}&B\Big([(x'', f''), (x, f)], (x', f')\Big) + B\Big((x, f), [(x'', f''), (x', f')]\Big) \\ &= B\Big(([x'', x], x''f - xf''), (x', f')\Big) + B\Big((x, f), ([x'', x'], x''f' - x'f'')\Big) \\ &= (x''f)(x') - (xf'')(x') + f'([x'', x]) + f([x'', x']) + (x''f')(x) - (x'f'')(x) \\ &= -f([x'', x']) + f''([x, x']) + f'([x'', x]) + f([x'', x']) \\ &\quad - f'([x'', x]) + f''([x', x]) \\ &= 0.\end{aligned}$$

The point of this exercise is that Cartan's semisimplicity criterion becomes false if the condition that the Killing form is nondegenerate is weakened to the mere existence of a nondegenerate invariant symmetric bilinear form.

Solution to Exercise 6.5.5. On the basis e, h, f of $\mathfrak{sl}_2$, the trace form has the matrix

$$\begin{pmatrix} 0 & 0 & 1 \\ 0 & 2 & 0 \\ 1 & 0 & 0 \end{pmatrix},$$

so the dual basis is $f, \frac{1}{2}h, e$ and the Casimir operator is given by

$$\Omega_{V(m)}(v) = efv + \tfrac{1}{2}h^2 v + fev \quad \text{for all } v \in V(m).$$

Applying this to a highest-weight vector w_0 of weight m and recalling the string formulas, we find that

$$\Omega_{V(m)}(w_0) = mw_0 + \tfrac{1}{2}m^2 w_0 + 0,$$

so $\Omega_{V(m)}$ must be multiplication by $m + \frac{1}{2}m^2$. For some unnecessary verification, let us compute $\Omega_{V(m)}(w_i)$ using the string formulas, without worrying about the boundary cases $i = 0$ and $i = m$:

$$\begin{aligned}\Omega_{V(m)}(w_i) &= efw_i + \tfrac{1}{2}h^2 w_i + few_i \\ &= (i+1)ew_{i+1} + \tfrac{1}{2}(m-2i)^2 w_i + (m-i+1)fw_{i-1} \\ &= (i+1)(m-i)w_i + (\tfrac{1}{2}m^2 - 2i(m-i))w_i + i(m-i+1)w_i \\ &= (m + \tfrac{1}{2}m^2)w_i.\end{aligned}$$

Solution to Exercise 6.5.6. Write v_0, v_1, v_2, v_3 for a string basis of $V(3)$. Recall from Example 6.1.10 that $v_i \otimes v_j$ is a weight vector of weight $6 - 2i - 2j$. Thus $\mathrm{Sym}^2(V(3))$ has the following basis of weight vectors:

$$\begin{aligned}&v_0 \otimes v_0 \quad \text{of weight } 6,\\ &v_1 \otimes v_1 \quad \text{of weight } 2,\\ &v_2 \otimes v_2 \quad \text{of weight } -2,\\ &v_3 \otimes v_3 \quad \text{of weight } -6,\\ &v_0 \otimes v_1 + v_1 \otimes v_0 \quad \text{of weight } 4,\\ &v_0 \otimes v_2 + v_2 \otimes v_0 \quad \text{of weight } 2,\\ &v_0 \otimes v_3 + v_3 \otimes v_0 \quad \text{of weight } 0,\\ &v_1 \otimes v_2 + v_2 \otimes v_1 \quad \text{of weight } 0,\\ &v_1 \otimes v_3 + v_3 \otimes v_1 \quad \text{of weight } -2,\\ &v_2 \otimes v_3 + v_3 \otimes v_2 \quad \text{of weight } -4.\end{aligned}$$

Thus the weights of the $\mathfrak{sl}_2$-module $\mathrm{Sym}^2(V(3))$, repeated according to their multiplicities, are $6, 4, 2, 2, 0, 0, -2, -2, -4, -6$. By Corollary 5.2.3(3) we deduce that $\mathrm{Sym}^2(V(3)) \cong V(6) \oplus V(2)$.

We can find the weights of $\mathrm{Sym}^3(V(2))$ similarly. Let w_0, w_1, w_2 be a string basis of $V(2)$. It is easy to see that $w_i \otimes w_j \otimes w_k$ is a weight vector of weight $6-2i-2j-2k$. Thus $\mathrm{Sym}^3(V(2))$ has the following basis of weight vectors:

$$\begin{aligned}&w_0 \otimes w_0 \otimes w_0 \quad \text{of weight } 6,\\ &w_1 \otimes w_1 \otimes w_1 \quad \text{of weight } 0,\\ &w_2 \otimes w_2 \otimes w_2 \quad \text{of weight } -6,\\ &w_0 \otimes w_0 \otimes w_1 + \text{sym. terms} \quad \text{of weight } 4,\\ &w_0 \otimes w_0 \otimes w_2 + \text{sym. terms} \quad \text{of weight } 2,\\ &w_1 \otimes w_1 \otimes w_0 + \text{sym. terms} \quad \text{of weight } 2,\end{aligned}$$

$$\begin{aligned}
&w_1 \otimes w_1 \otimes w_2 + \text{sym. terms} && \text{of weight } -2,\\
&w_2 \otimes w_2 \otimes w_0 + \text{sym. terms} && \text{of weight } -2,\\
&w_2 \otimes w_2 \otimes w_1 + \text{sym. terms} && \text{of weight } -4,\\
&w_0 \otimes w_1 \otimes w_2 + \text{sym. terms} && \text{of weight } 0;
\end{aligned}$$

here 'sym. terms' indicates that we add all tensors that can be obtained from the first term by permuting its factors. We deduce from the weights that $\mathrm{Sym}^3(V(2)) \cong V(6) \oplus V(2)$ also, as required. In fact, one can prove by a similar calculation that $\mathrm{Sym}^n(V(m)) \cong \mathrm{Sym}^m(V(n))$ for any $m, n \in \mathbb{N}$.

Solution to Exercise 6.5.7. For part (i), note that for $\left(\begin{smallmatrix} a & b \\ c & -a \end{smallmatrix}\right) \in \mathfrak{sl}_2$, $\left(\begin{smallmatrix} a_1 \\ a_2 \end{smallmatrix}\right), \left(\begin{smallmatrix} b_1 \\ b_2 \end{smallmatrix}\right) \in \mathbb{C}^2$, we have

$$\begin{aligned}
&B_2\left(\begin{pmatrix} a & b \\ c & -a \end{pmatrix}\begin{pmatrix} a_1 \\ a_2 \end{pmatrix}, \begin{pmatrix} b_1 \\ b_2 \end{pmatrix}\right) + B_2\left(\begin{pmatrix} a_1 \\ a_2 \end{pmatrix}, \begin{pmatrix} a & b \\ c & -a \end{pmatrix}\begin{pmatrix} b_1 \\ b_2 \end{pmatrix}\right)\\
&= B_2\left(\begin{pmatrix} aa_1 + ba_2 \\ ca_1 - aa_2 \end{pmatrix}, \begin{pmatrix} b_1 \\ b_2 \end{pmatrix}\right) + B_2\left(\begin{pmatrix} a_1 \\ a_2 \end{pmatrix}, \begin{pmatrix} ab_1 + bb_2 \\ cb_1 - ab_2 \end{pmatrix}\right)\\
&= (aa_1 + ba_2)b_2 - (ca_1 - aa_2)b_1 + a_1(cb_1 - ab_2) - a_2(ab_1 + bb_2) = 0.
\end{aligned}$$

So B_2 is $\mathfrak{sl}_2$-invariant.

For part (ii), recall that $\mathbb{C}^2 \otimes \mathbb{C}^2$ has the basis $v_1 \otimes v_1, v_1 \otimes v_2, v_2 \otimes v_1, v_2 \otimes v_2$. We can define $B(v_i \otimes w_j, v_{i'} \otimes w_{j'})$ by the formula given in the exercise; then, there is a unique way to extend this to a bilinear form B on the whole of $\mathbb{C}^2 \otimes \mathbb{C}^2$. Since the expressions $B(v \otimes w, v' \otimes w')$ and $B_2(v, v')B_2(w, w')$ are both linear in each of v, w, v', w' and are equal when v, w, v', w' belong to the basis of $\mathbb{C}^2$, they must be equal in general. To show that B is symmetric, it suffices to show that $B(v' \otimes w', v \otimes w) = B(v \otimes w, v' \otimes w')$ for all $v, w, v', w' \in \mathbb{C}^2$, which follows from the skew-symmetry of B_2:

$$B_2(v', v)B_2(w', w) = (-B_2(v, v'))(-B_2(w, w')) = B_2(v, v')B_2(w, w').$$

An easy calculation shows that the matrix of B relative to the above basis of $\mathbb{C}^2 \otimes \mathbb{C}^2$ is

$$\begin{pmatrix} 0 & 0 & 0 & 1 \\ 0 & 0 & -1 & 0 \\ 0 & -1 & 0 & 0 \\ 1 & 0 & 0 & 0 \end{pmatrix}.$$

Since this matrix is invertible, B is nondegenerate. This proves (ii).

To prove (iii) it suffices to note that, for all $x, y \in \mathfrak{sl}_2$ and $v, w, v', w' \in \mathbb{C}^2$,

$$\begin{aligned} &B\Big((x,y)(v\otimes w), v'\otimes w'\Big) + B\Big(v\otimes w, (x,y)(v'\otimes w')\Big) \\ &= B(xv\otimes w, v'\otimes w') + B(v\otimes yw, v'\otimes w') \\ &\quad + B(v\otimes w, xv'\otimes w') + B(v\otimes w, v'\otimes yw') \\ &= \Big(B_2(xv,v') + B_2(v,xv')\Big)B_2(w,w') + B_2(v,v')\Big(B_2(yw,w') + B_2(w,yw')\Big) \\ &= 0 + 0 = 0. \end{aligned}$$

We now come to (iv). As shown in Proposition 6.2.6, there is some basis t_1, t_2, t_3, t_4 of $\mathbb{C}^2 \otimes \mathbb{C}^2$ for which $B(t_i, t_j) = \delta_{ij}$. Part (iii) tells us that, relative to this basis, the matrices representing the action of elements of $\mathfrak{sl}_2 \times \mathfrak{sl}_2$ lie in $\mathfrak{so}_4$. Hence we get a homomorphism $\pi : \mathfrak{sl}_2 \times \mathfrak{sl}_2 \to \mathfrak{so}_4$, and all that remains is to prove that π is an isomorphism. Since both Lie algebras are six-dimensional, it suffices to show that $\ker(\pi) = \{0\}$, i.e. the representation of $\mathfrak{sl}_2 \times \mathfrak{sl}_2$ on $\mathbb{C}^2 \otimes \mathbb{C}^2$ is faithful. One way to do this would be to find the ideals of $\mathfrak{sl}_2 \times \mathfrak{sl}_2$ and show that none of the nonzero ideals can equal $\ker(\pi)$. A more direct way is to calculate the representing matrices for the obvious basis of $\mathfrak{sl}_2 \times \mathfrak{sl}_2$, relative to the standard basis of $\mathbb{C}^2 \otimes \mathbb{C}^2$ rather than the unknown basis t_1, t_2, t_3, t_4:

$$(e,0): \begin{pmatrix} 0&0&1&0\\0&0&0&1\\0&0&0&0\\0&0&0&0 \end{pmatrix}, \quad (h,0): \begin{pmatrix} 1&0&0&0\\0&1&0&0\\0&0&-1&0\\0&0&0&-1 \end{pmatrix}, \quad (f,0): \begin{pmatrix} 0&0&0&0\\0&0&0&0\\1&0&0&0\\0&1&0&0 \end{pmatrix},$$

$$(0,e): \begin{pmatrix} 0&1&0&0\\0&0&0&0\\0&0&0&1\\0&0&0&0 \end{pmatrix}, \quad (0,h): \begin{pmatrix} 1&0&0&0\\0&-1&0&0\\0&0&1&0\\0&0&0&-1 \end{pmatrix}, \quad (0,f): \begin{pmatrix} 0&0&0&0\\1&0&0&0\\0&0&0&0\\0&0&1&0 \end{pmatrix}.$$

These matrices are obviously linearly independent, so the representation is faithful.

Solution to Exercise 6.5.8. In this exercise it is convenient to use the notation $v \wedge w$ for $v \otimes w - w \otimes v \in \mathrm{Alt}^2(\mathbb{C}^4)$, for any $v, w \in \mathbb{C}^4$. Clearly $v \wedge v = 0$ and $w \wedge v = -(v \wedge w)$. Recall that $\mathrm{Alt}^2(\mathbb{C}^4)$ has a basis consisting of the elements $v_i \wedge v_j$, for $1 \leq i < j \leq n$.

For part (i), we can first define

$$B(v_i \wedge v_j, v_{i'} \wedge v_{j'}) = \det(v_i\ v_j\ v_{i'}\ v_{j'}) \quad \text{for } i < j \text{ and } i' < j'.$$

There is a unique way to extend B to a bilinear form on the whole of $\mathrm{Alt}^2(\mathbb{C}^4)$. Since the determinant of a matrix changes sign when you interchange two columns and is zero if two columns are equal, we see that

$$B(v_i \wedge v_j, v_{i'} \wedge v_{j'}) = \det(v_i\ v_j\ v_{i'}\ v_{j'}) \quad \text{for all } i, j, i', j'.$$

Then, since the expressions $B(v \wedge w, v' \wedge w')$ and $\det(v\ w\ v'\ w')$ are both linear in each of v, w, v', w' and are equal when v, w, v', w' belong to the basis of $\mathbb{C}^4$, they must be equal in general. To show that B is symmetric, it suffices to show that $B(v' \wedge w', v \wedge w) = B(v \wedge w, v' \wedge w')$ for all $v, w, v', w' \in \mathbb{C}^4$, which follows from the fact that

$$\det(v'\ w'\ v\ w) = -\det(v\ w'\ v'\ w) = \det(v\ w\ v'\ w').$$

An easy calculation shows that the matrix of B relative to the basis $v_1 \wedge v_2$, $v_1 \wedge v_3$, $v_1 \wedge v_4$, $v_2 \wedge v_3$, $v_2 \wedge v_4$, $v_3 \wedge v_4$ is

$$\begin{pmatrix} 0 & 0 & 0 & 0 & 0 & 1 \\ 0 & 0 & 0 & 0 & -1 & 0 \\ 0 & 0 & 0 & 1 & 0 & 0 \\ 0 & 0 & 1 & 0 & 0 & 0 \\ 0 & -1 & 0 & 0 & 0 & 0 \\ 1 & 0 & 0 & 0 & 0 & 0 \end{pmatrix}.$$

Since this matrix is invertible, B is nondegenerate. This proves (i).

To prove (ii), we note that, for all $x \in \mathfrak{sl}_4$ and $i, j, k, l \in \{1, 2, 3, 4\}$,

$$\begin{aligned} &B\Big(x(v_i \wedge v_j), v_k \wedge v_l\Big) + B\Big(v_i \wedge v_j, x(v_k \wedge v_l)\Big) \\ &= B(xv_i \wedge v_j, v_k \wedge v_l) + B(v_i \wedge xv_j, v_k \wedge v_l) \\ &\quad + B(v_i \wedge v_j, xv_k \wedge v_l) + B(v_i \wedge v_j, v_k \wedge xv_l) \\ &= \det(xv_i\ v_j\ v_k\ v_l) + \det(v_i\ xv_j\ v_k\ v_l) \\ &\quad + \det(v_i\ v_j\ xv_k\ v_l) + \det(v_i\ v_j\ v_k\ xv_l). \end{aligned}$$

We need to show that this sum of determinants is always zero. If i, j, k, l are not distinct, we may assume that $i = j$; then the third and fourth determinants will vanish (because in these matrices their first and second columns are the same), and the first and second determinants will be negatives of each other (because these matrices are related by interchange of the first two columns), so the sum is zero as required. If i, j, k, l are distinct, we may assume that $i = 1, j = 2, k = 3, l = 4$; then the four determinants are exactly the four diagonal entries of x, whose sum is zero because $x \in \mathfrak{sl}_4$. This proves (ii).

By Proposition 6.2.6, there is a basis $w_1, \ldots, w_6$ of $\mathrm{Alt}^2(\mathbb{C}^4)$ such that $B(w_i, w_j) = \delta_{ij}$ and, relative to this basis, the matrices representing $\mathfrak{sl}_4$ lie in $\mathfrak{so}_6$. Hence we have a Lie algebra homomorphism $\pi : \mathfrak{sl}_4 \to \mathfrak{so}_6$. We want to show that π is an isomorphism. Since both Lie algebras are 15-dimensional, it suffices to show that $\ker(\pi) = \{0\}$. But $\ker(\pi)$ is an ideal of $\mathfrak{sl}_4$, which is not the whole of $\mathfrak{sl}_4$ because the action of $\mathfrak{sl}_4$ on $\mathrm{Alt}^2(\mathbb{C}^4)$ is certainly not trivial. Since $\mathfrak{sl}_4$ is simple, we have $\ker(\pi) = \{0\}$ as required. Hence $\mathfrak{sl}_4 \cong \mathfrak{so}_6$.

Solutions for Chapter 7 exercises

Solution to Exercise 7.6.1. By Propositions 7.3.2 and 7.4.7 we need only show that $c_\lambda = c_{\lambda^*}$ for every $\lambda \in \Lambda^+$. Setting $\lambda = a_1\varepsilon_1 + \cdots + a_n\varepsilon_n$, so that $\lambda^* = -a_n\varepsilon_1 - \cdots - a_1\varepsilon_n$, we have

$$c_{\lambda^*} = \sum_{i=1}^{n} a_{n+1-i}^2 - (n+1-2i)a_{n+1-i} = \sum_{j=1}^{n} a_j^2 - (2j-(n+1))a_j = c_\lambda.$$

Solution to Exercise 7.6.2. We know that $\mathrm{Alt}^k(\mathbb{C}^n) \cong V(\varepsilon_1 + \cdots + \varepsilon_k)$. Applying Proposition 7.4.7 we deduce that

$$\mathrm{Alt}^k(\mathbb{C}^n)^* \cong V(-\varepsilon_{n+1-k} - \cdots - \varepsilon_n),$$
$$\mathrm{Alt}^k(\mathbb{C}^n)^* \otimes \mathbb{C}_{\mathrm{tr}} \cong V(\varepsilon_1 + \cdots + \varepsilon_{n-k}).$$

Hence $\mathrm{Alt}^k(\mathbb{C}^n)^* \otimes \mathbb{C}_{\mathrm{tr}} \cong \mathrm{Alt}^{n-k}(\mathbb{C}^n)$.

Using the $\mathfrak{gl}_n$-module isomorphism $\mathrm{Alt}^n(\mathbb{C}^n) \cong \mathbb{C}_{\mathrm{tr}}$ mentioned after Proposition 7.4.5 and the general isomorphism $V^* \otimes W \cong \mathrm{Hom}(V, W)$, we can reformulate the result of this exercise as $\mathrm{Alt}^{n-k}(\mathbb{C}^n) \cong \mathrm{Hom}(\mathrm{Alt}^k(\mathbb{C}^n), \mathrm{Alt}^n(\mathbb{C}^n))$. The isomorphism $\tau : \mathrm{Alt}^{n-k}(\mathbb{C}^n) \xrightarrow{\sim} \mathrm{Hom}(\mathrm{Alt}^k(\mathbb{C}^n), \mathrm{Alt}^n(\mathbb{C}^n))$ gives rise to a bilinear map $\beta : \mathrm{Alt}^{n-k}(\mathbb{C}^n) \times \mathrm{Alt}^k(\mathbb{C}^n) \to \mathrm{Alt}^n(\mathbb{C}^n)$, by the rule $\beta(x, y) = \tau(x)(y)$. This map β is the 'wedge product' of exterior algebra.

Solution to Exercise 7.6.3. If $a\varepsilon_1 + b\varepsilon_2 + c\varepsilon_3$ is a weight of $V(3\varepsilon_1 - \varepsilon_3)$ then we have

$$a \leq 3, \quad a + b \leq 3, \quad a + b + c = 2,$$

by Propositions 7.2.2 and 7.2.11(2). Moreover the same conditions must hold for any permutation of (a, b, c), by Proposition 7.1.16. It is easy to see that the set of weights that this allows is the union of the S_3-orbits of the following dominant weights:

$$3\varepsilon_1 - \varepsilon_3, \quad 2\varepsilon_1 + \varepsilon_2 - \varepsilon_3, \quad 2\varepsilon_1, \quad \varepsilon_1 + \varepsilon_2.$$

Again by Proposition 7.1.16, we need find the multiplicities of only these dominant weights. The multiplicity of the highest weight $3\varepsilon_1 - \varepsilon_3$ is 1, since the corresponding weight space is spanned by the highest-weight vector v. For each of the other weights, we determine which vectors $f_{i_1} f_{i_2} \cdots f_{i_r} v$ have that weight, with the following results:

$$2\varepsilon_1 + \varepsilon_2 - \varepsilon_3, \quad f_1 v;$$
$$2\varepsilon_1, \quad f_1 f_2 v,\ f_2 f_1 v;$$
$$\varepsilon_1 + \varepsilon_2, \quad f_1^2 f_2 v,\ f_1 f_2 f_1 v,\ f_2 f_1^2 v.$$

It remains to find the dimension of the span of each of these collections of vectors. Following Example 7.4.8, we can do this by constructing $V(3\varepsilon_1 - \varepsilon_3)$ as the submodule of $\mathrm{Sym}^3(\mathbb{C}^3) \otimes (\mathbb{C}^3)^*$ generated by the highest-weight vector $v = t_{(3,0,0)} \otimes v_3^*$. We obtain

$$\begin{aligned} f_1 v &= t_{(2,1,0)} \otimes v_3^*, \\ f_2 v &= -t_{(3,0,0)} \otimes v_2^*, \\ f_1^2 v &= 2t_{(1,2,0)} \otimes v_3^*, \\ f_1 f_2 v &= -t_{(2,1,0)} \otimes v_2^* + t_{(3,0,0)} \otimes v_1^*, \\ f_2 f_1 v &= t_{(2,0,1)} \otimes v_3^* - t_{(2,1,0)} \otimes v_2^*, \\ f_1^2 f_2 v &= -2t_{(1,2,0)} \otimes v_2^* + 2t_{(2,1,0)} \otimes v_1^*, \\ f_1 f_2 f_1 v &= t_{(1,1,1)} \otimes v_3^* - 2t_{(1,2,0)} \otimes v_2^* + t_{(2,1,0)} \otimes v_1^*, \\ f_2 f_1^2 v &= 2t_{(1,1,1)} \otimes v_3^* - 2t_{(1,2,0)} \otimes v_2^*. \end{aligned}$$

These calculations show that $f_1 v \neq 0$, so the multiplicity of the weight $2\varepsilon_1 + \varepsilon_2 - \varepsilon_3$ is 1, and that $f_1 f_2 v$ and $f_2 f_1 v$ are linearly independent, so the multiplicity of the weight $2\varepsilon_1$ is 2. In the remaining weight space we have a unique (up to a scalar) linear dependence relation:

$$f_1^2 f_2 v - 2 f_1 f_2 f_1 v + f_2 f_1^2 v = 0,$$

implying that the multiplicity of the weight $\varepsilon_1 + \varepsilon_2$ is 2. As a consequence, $\dim V(3\varepsilon_1 - \varepsilon_3) = 6 \times 1 + 6 \times 1 + 3 \times 2 + 3 \times 2 = 24$.

We could have predicted the linear dependence relation as follows. Since $[f_1, [f_1, f_2]] = 0$ in $\mathfrak{gl}_3$, for any vector v in a $\mathfrak{gl}_3$-module V we have

$$0 = [f_1, [f_1, f_2]]v = f_1[f_1, f_2]v - [f_1, f_2]f_1 v = f_1^2 f_2 v - 2 f_1 f_2 f_1 v + f_2 f_1^2 v.$$

Now Proposition 7.5.9 implies that we have a direct sum decomposition

$$\mathrm{Sym}^3(\mathbb{C}^3) \otimes (\mathbb{C}^3)^* = V(3\varepsilon_1 - \varepsilon_3) \oplus W',$$

where $W' \cong \mathrm{Sym}^2(\mathbb{C}^3)$. Hence, for any integral weight μ, the multiplicity of μ in $V(3\varepsilon_1 - \varepsilon_3)$ must equal the difference between its multiplicities in $\mathrm{Sym}^3(\mathbb{C}^3) \otimes (\mathbb{C}^3)^*$ and in $\mathrm{Sym}^2(\mathbb{C}^3)$. The latter multiplicities are easy to work out, because we have standard bases of $\mathrm{Sym}^3(\mathbb{C}^3) \otimes (\mathbb{C}^3)^*$ and of $\mathrm{Sym}^2(\mathbb{C}^3)$ that consist of weight vectors, so we just have to count how many basis vectors have weight μ. In $\mathrm{Sym}^3(\mathbb{C}^3) \otimes (\mathbb{C}^3)^*$, the basis vectors whose weight is dominant are as follows:

$$\begin{aligned} 3\varepsilon_1 - \varepsilon_3, &\quad t_{(3,0,0)} \otimes v_3^*; \\ 2\varepsilon_1 + \varepsilon_2 - \varepsilon_3, &\quad t_{(2,1,0)} \otimes v_3^*; \\ 2\varepsilon_1, &\quad t_{(2,0,1)} \otimes v_3^*, \ t_{(2,1,0)} \otimes v_2^*, \ t_{(3,0,0)} \otimes v_1^*; \\ \varepsilon_1 + \varepsilon_2, &\quad t_{(1,1,1)} \otimes v_3^*, \ t_{(1,2,0)} \otimes v_2^*, \ t_{(2,1,0)} \otimes v_1^*. \end{aligned}$$

In $\mathrm{Sym}^2(\mathbb{C}^3)$, the basis vectors whose weight is dominant are given by

$$2\varepsilon_1, \quad t_{(2,0,0)};$$
$$\varepsilon_1 + \varepsilon_2, \quad t_{(1,1,0)}.$$

Hence we recover the weight multiplicities found above.

Solution to Exercise 7.6.4. Let W' be a $\mathfrak{gl}_3$-submodule of $\mathrm{Sym}^2(\mathbb{C}^3) \otimes (\mathbb{C}^3)^*$ that is isomorphic to $\mathbb{C}^3$. We know that W' must be generated by a highest-weight vector w of weight ε_1. We also know, from the decomposition $\mathrm{Sym}^2(\mathbb{C}^3) \otimes (\mathbb{C}^3)^* = V(2\varepsilon_1 - \varepsilon_3) \oplus W'$, that w is unique up to a scalar; this means that W' is unique. The ε_1-weight space of $\mathrm{Sym}^2(\mathbb{C}^3) \otimes (\mathbb{C}^3)^*$ is the span of the basis vectors $t_{(2,0,0)} \otimes v_1^*, t_{(1,1,0)} \otimes v_2^*, t_{(1,0,1)} \otimes v_3^*$. To find w, we have to solve the following system of equations for $a, b, c \in \mathbb{C}$:

$$e_1(at_{(2,0,0)} \otimes v_1^* + bt_{(1,1,0)} \otimes v_2^* + ct_{(1,0,1)} \otimes v_3^*) = 0,$$
$$e_2(at_{(2,0,0)} \otimes v_1^* + bt_{(1,1,0)} \otimes v_2^* + ct_{(1,0,1)} \otimes v_3^*) = 0.$$

Now some easy calculations show that

$$e_1(t_{(2,0,0)} \otimes v_1^*) = -t_{(2,0,0)} \otimes v_2^*,$$
$$e_1(t_{(1,1,0)} \otimes v_2^*) = 2t_{(2,0,0)} \otimes v_2^*,$$
$$e_1(t_{(1,0,1)} \otimes v_3^*) = 0,$$
$$e_2(t_{(2,0,0)} \otimes v_1^*) = 0,$$
$$e_2(t_{(1,1,0)} \otimes v_2^*) = -t_{(1,1,0)} \otimes v_3^*,$$
$$e_2(t_{(1,0,1)} \otimes v_3^*) = t_{(1,1,0)} \otimes v_3^*.$$

So the above system of equations reduces to $-a + 2b = 0$ and $-b + c = 0$, giving the (unique up to a scalar) solution $a = 2, b = 1, c = 1$. So W' has a basis consisting of the following elements:

$$w = 2t_{(2,0,0)} \otimes v_1^* + t_{(1,1,0)} \otimes v_2^* + t_{(1,0,1)} \otimes v_3^*,$$
$$f_1 w = t_{(1,1,0)} \otimes v_1^* + 2t_{(0,2,0)} \otimes v_2^* + t_{(0,1,1)} \otimes v_3^*,$$
$$f_2 f_1 w = t_{(1,0,1)} \otimes v_1^* + t_{(0,1,1)} \otimes v_2^* + 2t_{(0,0,2)} \otimes v_3^*.$$

Solution to Exercise 7.6.5. The assumption that $\dim V_\lambda = \dim W_\lambda$ for all $\lambda \in \Lambda$ clearly implies that $\dim V = \dim W$. We prove the claim by induction on $\dim V$. When $\dim V = 0$ the statement is trivial, so we assume that $\dim V \geq 1$ and that the statement is known for integral $\mathfrak{gl}_n$-modules of smaller dimension. Let X be the set of weights of V, which by assumption is also the set of weights of W. Since X is finite we can find some $\lambda \in X$ which is maximal for the partial order $\leq$, in the sense that if $\lambda < \mu$ then $\mu \notin X$. Then any weight vector in V of weight λ is a highest-weight vector, and similarly for W. This implies that $\lambda \in \Lambda^+$.

By Corollary 7.5.5 we have direct sum decompositions $V = V_{[\lambda]} \oplus V'$ and $W = W_{[\lambda]} \oplus W'$, where $V_{[\lambda]}$ and $W_{[\lambda]}$ are direct sums of irreducible submodules isomorphic to $V(\lambda)$ and V' and W' are direct sums of irreducible submodules not isomorphic to $V(\lambda)$. Moreover, when $V_{[\lambda]}$ is written as a direct sum of copies of $V(\lambda)$, the number of summands equals $\dim V_\lambda \cap V^{\mathfrak{n}^+} = \dim V_\lambda$ and similarly for $W_{[\lambda]}$. By assumption, $\dim V_\lambda = \dim W_\lambda$ so $V_{[\lambda]} \cong W_{[\lambda]}$. It follows that V' and W' have the same weight multiplicities, so by the induction hypothesis we have $V' \cong W'$. Hence $V \cong W$ as required.

Solution to Exercise 7.6.6. By Proposition 7.5.8, we have

$$c_\lambda + c_\mu + \frac{2}{n}\lambda(1_n)\mu(1_n) \le c_{\lambda+\mu},$$

with equality if and only if $V(\lambda)\otimes V(\mu)$ is irreducible. If we let $\lambda = a_1\varepsilon_1 + \cdots + a_n\varepsilon_n$ and $\mu = b_1\varepsilon_1 + \cdots + b_n\varepsilon_n$ and make some obvious cancellations, the above inequality becomes

$$(a_1 + \cdots + a_n)(b_1 + \cdots + b_n) \le n(a_1b_1 + \cdots + a_nb_n),$$

which is known as Chebyshev's inequality; it holds for any real numbers a_i, b_i under the assumption that $a_1 \ge \cdots \ge a_n$ and $b_1 \ge \cdots \ge b_n$. The proof is as follows:

$$0 \le \sum_{i,j}(a_i - a_j)(b_i - b_j) = 2n\sum_i a_ib_i - 2\left(\sum_i a_i\right)\left(\sum_i b_i\right).$$

In particular, equality holds in Chebyshev's inequality if and only if, for every $i < j$, either $a_i = a_j$ or $b_i = b_j$. Considering the case $i = 1, j = n$, we see that this condition is equivalent to the condition that either $a_1 = \cdots = a_n$ or $b_1 = \cdots = b_n$. So $V(\lambda) \otimes V(\mu)$ is irreducible only in the trivial cases when either $V(\lambda)$ or $V(\mu)$ is one-dimensional.

Solution to Exercise 7.6.7. Since the map $\mathbb{C}^n \otimes (\mathbb{C}^n)^* \to \mathbb{C}$ that sends $v_i \otimes v_j^*$ to δ_{ij} is a $\mathfrak{gl}_n$-module homomorphism, so is the linear map $\psi : (\mathbb{C}^n)^{\otimes k} \otimes (\mathbb{C}^n)^* \to (\mathbb{C}^n)^{\otimes k-1}$ defined by

$$\psi(v_{i_1} \otimes \cdots \otimes v_{i_k} \otimes v_j^*) = \frac{1}{k}\sum_{a=1}^{k}(-1)^a \delta_{i_a j}\, v_{i_1} \otimes \cdots \otimes \widehat{v_{i_a}} \otimes \cdots \otimes v_{i_k},$$

where as before the caret indicates that the corresponding factor is not present. To work out the restriction of ψ to the submodule $\mathrm{Alt}^k(\mathbb{C}^n) \otimes (\mathbb{C}^n)^*$, we must evaluate

$\psi(u_{i_1,\dots,i_k} \otimes v_j^*)$. Clearly this is zero if $j \notin \{i_1, \dots, i_k\}$. If $j = i_b$ for some $b \in \{1, \dots, k\}$ then we have

$$\begin{aligned}\psi(u_{i_1,\dots,i_k} \otimes v_{i_b}^*) &= \sum_{\sigma \in S_k} \varepsilon(\sigma)\psi(v_{i_{\sigma(1)}} \otimes \cdots \otimes v_{i_{\sigma(k)}} \otimes v_{i_b}^*) \\ &= \frac{1}{k} \sum_{\sigma \in S_k} \varepsilon(\sigma)(-1)^{\sigma^{-1}(b)} v_{i_{\sigma(1)}} \otimes \cdots \otimes \widehat{v_{i_b}} \otimes \cdots \otimes v_{i_{\sigma(k)}}.\end{aligned}$$

The terms in this last expression involve products of the same $k-1$ vectors, namely $v_{i_1}, \dots, \widehat{v_{i_b}}, \dots, v_{i_k}$, but in various different orders. Call these vectors $v_{j_1}, \dots, v_{j_{k-1}}$, so that $j_s = i_s$ for $s < b$ and $j_s = i_{s+1}$ for $s \geq b$. For any $\tau \in S_{k-1}$, the terms that involve the product $v_{j_{\tau(1)}} \otimes \cdots \otimes v_{j_{\tau(k-1)}}$ are those indexed by $\sigma \in S_k$ with the property that the k-tuple $(\sigma(1), \dots, \sigma(k))$ is obtained from the $(k-1)$-tuple $(\tau(1), \dots, \tau(k-1))$ by adding 1 to all entries that are at least b and also inserting an entry b at any of the k possible positions. Call these permutations $\sigma_1, \dots, \sigma_k$, where $\sigma_i(i) = b$. Then $\sigma_i = \sigma_1 c_i$, where c_i is the i-cycle $(1\,2\,\dots\,i)$, so $\varepsilon(\sigma_i) = \varepsilon(\sigma_1)(-1)^{i-1}$. Since $\varepsilon(\sigma_b) = \varepsilon(\tau)$ we get $\varepsilon(\sigma_i) = (-1)^{i-b}\varepsilon(\tau)$. Hence the coefficient of $v_{j_{\tau(1)}} \otimes \cdots \otimes v_{j_{\tau(k-1)}}$ in the above expression is $\frac{1}{k}\sum_{i=1}^k \varepsilon(\sigma_i)(-1)^i = (-1)^b\varepsilon(\tau)$. Consequently, we have

$$\psi(u_{i_1,\dots,i_k} \otimes v_j^*) = \begin{cases} (-1)^b\, u_{i_1,\dots,\widehat{i_b},\dots,i_k} & \text{if } j = i_b \text{ for some } b, \\ 0 & \text{otherwise.} \end{cases}$$

In particular, the image of ψ restricted to $\mathrm{Alt}^k(\mathbb{C}^n) \otimes (\mathbb{C}^n)^*$ is exactly $\mathrm{Alt}^{k-1}(\mathbb{C}^n)$. This restriction thus gives us a surjective $\mathfrak{gl}_n$-module homomorphism φ : $\mathrm{Alt}^k(\mathbb{C}^n) \otimes (\mathbb{C}^n)^* \to \mathrm{Alt}^{k-1}(\mathbb{C}^n)$.

We now imitate the corresponding argument in the proof of Proposition 7.5.9. Let $V = \mathrm{Alt}^k(\mathbb{C}^n) \otimes (\mathbb{C}^n)^*$, $W = \ker(\varphi)$. Since $V/W \cong \mathrm{Alt}^{k-1}(\mathbb{C}^n)$, the Casimir operator $\Omega_{V/W}$ is scalar multiplication by $c_{\varepsilon_1+\cdots+\varepsilon_{k-1}}$. On the one hand, by Proposition 7.5.8(2) every eigenvalue of Ω_V is $\leq c_{\varepsilon_1+\cdots+\varepsilon_k-\varepsilon_n}$. Hence

$$\begin{aligned}\mathrm{tr}(\Omega_V) &\leq c_{\varepsilon_1+\cdots+\varepsilon_k-\varepsilon_n} \dim W + c_{\varepsilon_1+\cdots+\varepsilon_{k-1}} \dim V/W \\ &= c_{\varepsilon_1+\cdots+\varepsilon_k-\varepsilon_n} \dim V - (c_{\varepsilon_1+\cdots+\varepsilon_k-\varepsilon_n} - c_{\varepsilon_1+\cdots+\varepsilon_{k-1}}) \dim V/W \\ &= (k(n-k)+n+k)\dim V - 2(n-k+1)\binom{n}{k-1} \\ &= (c_{\varepsilon_1+\cdots+\varepsilon_k} + c_{-\varepsilon_n})\dim V - 2k\binom{n}{k} \\ &= \left(c_{\varepsilon_1+\cdots+\varepsilon_k} + c_{-\varepsilon_n} - \frac{2k}{n}\right)\dim V.\end{aligned}$$

On the other hand, Proposition 7.5.8(4) tells us that $\mathrm{tr}(\Omega_V)$ equals this last quantity. We conclude that the $c_{\varepsilon_1+\cdots+\varepsilon_k-\varepsilon_n}$-eigenspace of Ω_V equals W, so $W \cong V(\varepsilon_1 + \cdots + \varepsilon_k - \varepsilon_n)$ by Proposition 7.5.8(3). Hence $V(\varepsilon_1 + \cdots + \varepsilon_k - \varepsilon_n)$ can be described as the kernel of the surjective $\mathfrak{gl}_n$-module homomorphism $\varphi : \mathrm{Alt}^k(\mathbb{C}^n) \otimes (\mathbb{C}^n)^* \to \mathrm{Alt}^{k-1}(\mathbb{C}^n)$.

Solution to Exercise 7.6.8. We have an obvious direct sum decomposition

$$\mathbb{C}^n \otimes \mathbb{C}^n \otimes \mathbb{C}^n = \mathrm{Sym}^2(\mathbb{C}^n) \otimes \mathbb{C}^n \oplus \mathrm{Alt}^2(\mathbb{C}^n) \otimes \mathbb{C}^n.$$

We also know two irreducible submodules of $\mathbb{C}^n \otimes \mathbb{C}^n \otimes \mathbb{C}^n$, namely $\mathrm{Sym}^3(\mathbb{C}^n)$ and $\mathrm{Alt}^3(\mathbb{C}^n)$. (Since $\mathrm{Alt}^3(\mathbb{C}^2) = \{0\}$, we must ignore $\mathrm{Alt}^3(\mathbb{C}^n)$ in the $n = 2$ case – this applies throughout the solution.) From the relevant definitions it is clear that $\mathrm{Sym}^3(\mathbb{C}^n)$ is a submodule of $\mathrm{Sym}^2(\mathbb{C}^n) \otimes \mathbb{C}^n$ and $\mathrm{Alt}^3(\mathbb{C}^n)$ is a submodule of $\mathrm{Alt}^2(\mathbb{C}^n) \otimes \mathbb{C}^n$. By complete reducibility (see Corollary 7.5.4), we have direct sum decompositions

$$\mathrm{Sym}^2(\mathbb{C}^n) \otimes \mathbb{C}^n = \mathrm{Sym}^3(\mathbb{C}^n) \oplus W \quad \text{and} \quad \mathrm{Alt}^2(\mathbb{C}^n) \otimes \mathbb{C}^n = \mathrm{Alt}^3(\mathbb{C}^n) \oplus W'.$$

We can easily find the weight multiplicities of W and W'. The weights of $\mathrm{Sym}^2(\mathbb{C}^n)$ are $\varepsilon_i + \varepsilon_j$ for $1 \le i \le j \le n$, each with multiplicity 1. Hence the weights of $\mathrm{Sym}^2(\mathbb{C}^n) \otimes \mathbb{C}^n$ are: $3\varepsilon_i$ for $1 \le i \le n$ with multiplicity 1; $2\varepsilon_i + \varepsilon_j$ and $\varepsilon_i + 2\varepsilon_j$ for $1 \le i < j \le n$ with multiplicity 2; and $\varepsilon_i + \varepsilon_j + \varepsilon_k$ for $1 \le i < j < k \le n$ with multiplicity 3. Subtracting the weight multiplicities of $\mathrm{Sym}^3(\mathbb{C}^n)$ we find that the weights of W are: $2\varepsilon_i + \varepsilon_j$ and $\varepsilon_i + 2\varepsilon_j$ for $1 \le i < j \le n$ with multiplicity 1; $\varepsilon_i + \varepsilon_j + \varepsilon_k$ for $1 \le i < j < k \le n$ with multiplicity 2. A similar calculation shows that the weights of W', and their multiplicities, are the same as for W. By Exercise 7.6.5, we can conclude that $W' \cong W$.

To decompose W as a direct sum of irreducible modules, we note that among the weights of W there is one that is maximal for the partial order $\le$, namely $2\varepsilon_1 + \varepsilon_2$, whose multiplicity in W is 1. So W contains a unique (up to a scalar) highest-weight vector of weight $2\varepsilon_1 + \varepsilon_2$ and hence a unique submodule isomorphic to $V(2\varepsilon_1 + \varepsilon_2)$. The only other weight of W that is dominant is $\varepsilon_1 + \varepsilon_2 + \varepsilon_3$, whose multiplicity in W is 2. So we have

$$W \cong V(2\varepsilon_1 + \varepsilon_2) \oplus W'',$$

where W'' is the direct sum of s isomorphic copies of $\mathrm{Alt}^3(\mathbb{C}^n)$ for some $s \in \{0, 1, 2\}$. One way to find s would be to determine explicitly the highest-weight vectors in $\mathrm{Sym}^2(\mathbb{C}^n) \otimes \mathbb{C}^n$ of weight $\varepsilon_1 + \varepsilon_2 + \varepsilon_3$. A simpler way is to calculate the trace of the Casimir operator Ω_W using Proposition 7.5.8. We have on the one hand

$$\begin{aligned}
\operatorname{tr}(\Omega_W) &= \operatorname{tr}(\Omega_{\operatorname{Sym}^2(\mathbb{C}^n)\otimes\mathbb{C}^n}) - \operatorname{tr}(\Omega_{\operatorname{Sym}^3(\mathbb{C}^n)}) \\
&= n\binom{n+1}{2}\left(c_{2\varepsilon_1} + c_{\varepsilon_1} + \frac{4}{n}\right) - \binom{n+2}{3} c_{3\varepsilon_1} \\
&= n\binom{n+1}{2}\left(2n+2+n+\frac{4}{n}\right) - \binom{n+2}{3}(3n+6) \\
&= \frac{1}{2}n(n+1)\left(3n^2+2n+4-(n+2)^2\right) \\
&= n^2(n+1)(n-1)
\end{aligned}$$

and on the other hand

$$\begin{aligned}
\operatorname{tr}(\Omega_W) &= \operatorname{tr}(\Omega_{V(2\varepsilon_1+\varepsilon_2)}) + \operatorname{tr}(\Omega_{W''}) \\
&= c_{2\varepsilon_1+\varepsilon_2} \dim V(2\varepsilon_1+\varepsilon_2) + c_{\varepsilon_1+\varepsilon_2+\varepsilon_3} \dim W'' \\
&= 3n \dim V(2\varepsilon_1+\varepsilon_2) + (3n-6)\dim W'' \\
&= 3n \dim W - 6 \dim W'' \\
&= 3n\left(n\binom{n+1}{2} - \binom{n+2}{3}\right) - 6\dim W'' \\
&= n^2(n+1)(n-1) - 6\dim W'',
\end{aligned}$$

and we conclude that $W'' = \{0\}$. Hence $W \cong V(2\varepsilon_1+\varepsilon_2)$ is irreducible.

The conclusion is that we have the following decomposition of $\mathbb{C}^n \otimes \mathbb{C}^n \otimes \mathbb{C}^n$ as a direct sum of irreducible $\mathfrak{gl}_n$-submodules:

$$\mathbb{C}^n \otimes \mathbb{C}^n \otimes \mathbb{C}^n = \operatorname{Sym}^3(\mathbb{C}^n) \oplus W \oplus W' \oplus \operatorname{Alt}^3(\mathbb{C}^n),$$

where W and W' are both isomorphic to $V(2\varepsilon_1+\varepsilon_2)$. If we had instead used $\mathbb{C}^n \otimes \operatorname{Sym}^2(\mathbb{C}^n)$ and $\mathbb{C}^n \otimes \operatorname{Alt}^2(\mathbb{C}^n)$ as our initial submodules, we would have obtained not the same W and W' but rather two different submodules isomorphic to $V(2\varepsilon_1+\varepsilon_2)$. However, the sum $W \oplus W'$ would have been the same, because it is canonically defined as the submodule of $\mathbb{C}^n \otimes \mathbb{C}^n \otimes \mathbb{C}^n$ generated by the highest-weight vectors of weight $2\varepsilon_1+\varepsilon_2$.

Solution to Exercise 7.6.9

(i) It suffices to show that $\Psi_V(e_{ij}v) - e_{ij}\Psi_V(v) = 0$ for all $v \in V$ and all i, j. By definition, this difference equals

$$\sum_{i_1,i_2,i_3} (e_{i_1i_2}e_{i_2i_3}e_{i_3i_1}e_{ij}v - e_{ij}e_{i_1i_2}e_{i_2i_3}e_{i_3i_1}v).$$

Using the defining property of a $\mathfrak{gl}_3$-module repeatedly, we can rewrite this expression as

$$\sum_{i_1,i_2,i_3} e_{i_1 i_2} e_{i_2 i_3} [e_{i_3 i_1}, e_{ij}] v + \sum_{i_1,i_2,i_3} e_{i_1 i_2} [e_{i_2 i_3}, e_{ij}] e_{i_3 i_1} v$$
$$+ \sum_{i_1,i_2,i_3} [e_{i_1 i_2}, e_{ij}] e_{i_2 i_3} e_{i_3 i_1} v.$$

Now using the rule $[e_{kl}, e_{ij}] = \delta_{li} e_{kj} - \delta_{jk} e_{il}$, these three sums become, respectively,

$$\sum_{i_2,i_3} e_{i i_2} e_{i_2 i_3} e_{i_3 j} v - \sum_{i_1,i_2} e_{i_1 i_2} e_{i_2 j} e_{i i_1} v,$$
$$\sum_{i_1,i_2} e_{i_1 i_2} e_{i_2 j} e_{i i_1} v - \sum_{i_1,i_3} e_{i_1 j} e_{i i_3} e_{i_3 i_1} v,$$
$$\sum_{i_1,i_3} e_{i_1 j} e_{i i_3} e_{i_3 i_1} v - \sum_{i_2,i_3} e_{i i_2} e_{i_2 i_3} e_{i_3 j} v,$$

and the total is now clearly 0 as required.

(ii) It suffices to compute $\Psi_V(v)$, where v is a highest-weight vector of weight $a_1\varepsilon_1 + a_2\varepsilon_2 + a_3\varepsilon_3$. By definition this means that $e_{ij}v = 0$ if $i < j$ and $e_{ii}v = a_i v$. So the sum of the terms $e_{i_1 i_2} e_{i_2 i_3} e_{i_3 i_1} v$ where $i_3 < i_1$ is zero and the sum of the terms where $i_3 = i_1$ can be worked out in the same way as that in which we computed the scalar for the Casimir operator in Proposition 7.3.2:

$$\begin{aligned}
\sum_{i_1,i_2} e_{i_1 i_2} e_{i_2 i_1} e_{i_1 i_1} v &= \sum_{i_1,i_2} a_{i_1} e_{i_1 i_2} e_{i_2 i_1} v \\
&= \sum_{i_1} a_{i_1} e_{i_1 i_1} e_{i_1 i_1} v + \sum_{i_1 < i_2} a_{i_1} e_{i_1 i_2} e_{i_2 i_1} v \\
&= \sum_{i_1} a_{i_1}^3 v + \sum_{i_1 < i_2} a_{i_1} [e_{i_1 i_2}, e_{i_2 i_1}] v \\
&= (a_1^3 + a_2^3 + a_3^3) v + \sum_{i_1 < i_2} a_{i_1} (e_{i_1 i_1} v - e_{i_2 i_2} v) \\
&= (a_1^3 + a_2^3 + a_3^3 + 2a_1^2 + a_2^2 - a_1 a_2 - a_1 a_3 - a_2 a_3) v.
\end{aligned}$$

This leaves the terms $e_{i_1 i_2} e_{i_2 i_3} e_{i_3 i_1} v$ where $i_3 > i_1$. Of these, the three terms where $i_3 > i_2 = i_1$ have the following sum:

$$\begin{aligned}
\sum_{i_1 < i_3} e_{i_1 i_1} e_{i_1 i_3} e_{i_3 i_1} v &= \sum_{i_1 < i_3} e_{i_1 i_1} [e_{i_1 i_3}, e_{i_3 i_1}] v \\
&= \sum_{i_1 < i_3} e_{i_1 i_1} (e_{i_1 i_1} v - e_{i_3 i_3} v) \\
&= \sum_{i_1 < i_3} a_{i_1} (a_{i_1} - a_{i_3}) v \\
&= (2a_1^2 + a_2^2 - a_1 a_2 - a_1 a_3 - a_2 a_3) v.
\end{aligned}$$

The three terms where $i_2 = i_3 > i_1$ have the sum

$$\begin{aligned}\sum_{i_1<i_3} e_{i_1i_3}e_{i_3i_3}e_{i_3i_1}v &= \sum_{i_1<i_3}(e_{i_1i_3}[e_{i_3i_3}, e_{i_3i_1}]v + e_{i_1i_3}e_{i_3i_1}e_{i_3i_3}v)\\ &= \sum_{i_1<i_3}(1+a_{i_3})e_{i_1i_3}e_{i_3i_1}v\\ &= \sum_{i_1<i_3}(1+a_{i_3})(a_{i_1}-a_{i_3})v\\ &= (-a_2^2 - 2a_3^2 + a_1a_2 + a_1a_3 + a_2a_3 + 2a_1 - 2a_3)v.\end{aligned}$$

This leaves three terms, which we can work out separately:

$$\begin{aligned}e_{21}e_{13}e_{32}v &= e_{21}[e_{13}, e_{32}]v = e_{21}e_{12}v = 0,\\ e_{12}e_{23}e_{31}v &= e_{12}[e_{23}, e_{31}]v = e_{12}e_{21}v = (a_1 - a_2)v,\\ e_{13}e_{32}e_{21}v &= [e_{13}, e_{32}]e_{21}v + e_{32}e_{13}e_{21}v\\ &= e_{12}e_{21}v + e_{32}[e_{13}, e_{21}]v\\ &= (a_1 - a_2)v - e_{32}e_{23}v = (a_1 - a_2)v.\end{aligned}$$

Summing up, we conclude that the scalar is

$$a_1^3 + a_2^3 + a_3^3 + 4a_1^2 + a_2^2 - 2a_3^2 - a_1a_2 - a_1a_3 - a_2a_3 + 4a_1 - 2a_2 - 2a_3.$$

This scalar (like that for the Casimir operator for $\mathfrak{gl}_3$) is symmetric not in a_1, a_2, a_3, but in the shifted variables $a_1 + 1, a_2, a_3 - 1$.

One can define a degree-k generalized Casimir operator $\Psi_V^{(k)}$ on any $\mathfrak{gl}_n$-module by an obvious generalization of the definition of Ψ_V. The degree-1 operator is simply the action of 1_n on V, and the degree-2 operator is the ordinary Casimir operator. It is true in general that $\Psi_V^{(k)}$ is a $\mathfrak{gl}_n$-module endomorphism and that if $V \cong V(a_1\varepsilon_1 + \cdots + a_n\varepsilon_n)$ then $\Psi_V^{(k)}$ is multiplication by scalar which is a degree-k symmetric polynomial in $a_1 + \frac{1}{2}(n-1), a_2 + \frac{1}{2}(n-3), \ldots, a_{n-1} + \frac{1}{2}(3-n), a_n + \frac{1}{2}(1-n)$.

References

[1] N. BOURBAKI, *Lie Groups and Lie Algebras*, 3 vols., Springer-Verlag, 1989 (Chapters 1–3), 2002 (Chapters 4–6), 2005 (Chapters 7–9).

[2] R. W. CARTER, *Lie Algebras of Finite and Affine Type*, vol. 96 of Cambridge Studies in Advanced Mathematics, Cambridge University Press, 2005.

[3] R. W. CARTER, G. SEGAL, AND I. G. MACDONALD, *Lectures on Lie Groups and Lie Algebras*, vol. 32 of London Mathematical Society Student Texts, Cambridge University Press, 1995.

[4] K. ERDMANN AND M. J. WILDON, *Introduction to Lie Algebras*, Springer Undergraduate Mathematics Series, Springer-Verlag, 2006.

[5] W. FULTON, *Young Tableaux*, vol. 35 of London Mathematical Society Student Texts, Cambridge University Press, 1997.

[6] W. FULTON AND J. HARRIS, *Representation Theory: A First Course*, vol. 129 of Graduate Texts in Mathematics, Springer-Verlag, 1991.

[7] R. GOODMAN AND N. R. WALLACH, *Representations and Invariants of the Classical Groups*, vol. 68 of Encyclopedia of Mathematics and its Applications, Cambridge University Press, 1998.

[8] B. C. HALL, *Lie Groups, Lie Algebras, and Representations: An Elementary Introduction*, vol. 222 of Graduate Texts in Mathematics, Springer-Verlag, 2003.

[9] J. HONG AND S.-J. KANG, *Introduction to Quantum Groups and Crystal Bases*, vol. 42 of Graduate Studies in Mathematics, American Mathematical Society, 2002.

[10] J. E. HUMPHREYS, *Introduction to Lie Algebras and Representation Theory*, vol. 9 of Graduate Texts in Mathematics, Springer-Verlag, 1972.

[11] J. C. JANTZEN, *Lectures on Quantum Groups*, vol. 6 of Graduate Studies in Mathematics, American Mathematical Society, 1996.

[12] V. G. KAC, *Infinite-Dimensional Lie Algebras*, 3rd edn, Cambridge University Press, 1990.

[13] A. W. KNAPP, *Lie Groups: Beyond an Introduction*, 2nd edn, vol. 140 of Progress in Mathematics, Birkhäuser, 2002.

[14] I. G. MACDONALD, *Symmetric Functions and Hall Polynomials*, 2nd edn, Oxford University Press, 1995.

[15] W. ROSSMANN, *Lie Groups: An Introduction Through Linear Groups*, vol. 5 of Oxford Graduate Texts in Mathematics, Oxford University Press, 2002.

Index

action, 33–6
alternating tensor, 5–7, 64–5, 74, 94–7, 104, 110
associative algebra, 14, 31, 34–5
automorphism, 19

bilinear form, 66–7, 69–73, 76–7
  degenerate, 66–7, 69–73, 76–7
  invariant, 66–7, 69–70, 72–3, 76–7
  nondegenerate, 66–7, 69–73, 76–7
  skew-symmetric, 67, 69
  symmetric, 67, 69–70, 76–7
branching rule, 113

canonical basis, 113–14
Cartan matrix, 108
Cartan's semisimplicity criterion, 72, 76, 109
Cartan–Killing classification, 30, 106–8
Casimir operator, 72–6, 90–2, 99, 102–4, 109–10
Cayley–Dickson algebra, 109
centralizer, 32
centre, 25, 31–2, 68
  of a direct product, 30
  of a simple Lie algebra, 29
  of $\mathfrak{gl}_n$, 26, 32
character
  of a Lie algebra, 37–9, 41, 48, 61–2
  of a module, 111
commutator, 9, 13–5
crystal basis, 113–14

density theorem, 43
derivation, 22–3
derivative, 8–12
derived algebra, 25–6, 29–32
  of a direct product, 30
  of a simple Lie algebra, 29
  of $\mathfrak{gl}_n$, 26
determinant, 2, 6–7, 111
diagonal matrix, 15, 21, 27, 29, 78, 107
direct product, 30–1, 48
direct sum, 4, 27, 31–2, 44
dual basis, 3, 60, 72, 79
dual space, 3, 38, 60–1, 65–7
Dynkin diagram, 108

eigenspace, *see* eigenvector
eigenvalue, *see* eigenvector
eigenvector, 26, 42–3, 46, 48, 64, 68, 102–4
endomorphism, 38, 66, 68, 72, 105

fundamental homomorphism theorem, 28–30, 36

Gel'fand–Tsetlin basis, 113
group
  general linear, 1, 10
  of automorphisms, 19
  special linear, 11
  special orthogonal, 12

highest-weight vector
  for a simple Lie algebra, 109–10, 113
  for $\mathfrak{gl}_n$, 86–98, 101–4
  for $\mathfrak{sl}_2$, 52–6, 58, 61, 66, 76
homomorphism
  of Lie algebras, 10, 17, 19, 28–9
  of modules, 38, 41, 65–6
  of vector spaces, 65

ideal, 24–32
  abelian, 72, 76

nontrivial, 25
of $\mathfrak{gl}_n$, 26–7
of Mat_n, 26
invariant vector, 42, 65–6
isomorphism
of Lie algebras, 17–20, 28
of modules, 38–41, 46–8

Jacobi identity, 13–16, 18–20, 25, 37

Kac–Moody algebra, 109
Killing form, 70–2, 76, 109
Kostka number, 112

Lie algebra, 1, 10, 13–20
abelian, 15–19, 22, 25, 42, 46
general linear, 10, 14
linear, 14, 21
of derivations, 22, 31, 48, 109
of linear transformations, 22, 33
one-dimensional, 15, 17, 29, 46
semisimple, 108
simple, 29–30, 48, 106–14
special linear, 12, 14, 18–19, 30, 48
special orthogonal, 12, 14, 19, 30, 106
three-dimensional, 18–20, 30, 32, 48
two-dimensional, 16–17
Lie bracket, 13–20
of $\mathfrak{gl}_n$, 16
of ideals, 25–6
of $\mathfrak{sl}_2$, 18
of $\mathfrak{so}_3$, 19
Lie group, 1, 11, 106
Lie's theorem, 12

Maschke's theorem, 46
module, 33–49
completely reducible, 45–8, 55, 99–101
faithful, 34–7
highest-weight, 52–5, 87–91, 109–11
infinite-dimensional, 59, 111
integral, 47, 79–83, 86–7, 90–3, 96, 100–2, 104
irreducible, 41–8, 54, 91, 96–9, 109–11
natural, 35–6, 43, 50
one-dimensional, 37–9, 41
over a simple Lie algebra, 109–14
over $\mathfrak{b}_n$, 43, 45
over $\mathfrak{d}_n$, 45
over $\mathfrak{gl}_n$, 35, 43, 69, 78–105, 111–13
over $\mathfrak{l}_2$, 34–5, 39–41, 47, 58, 62, 64–5
over Mat_n, 35
over $\mathfrak{sl}_2$, 34, 37, 47, 50–9, 61, 63, 65–6, 76, 83, 94, 99
over $\mathfrak{sl}_n$, 43, 69, 82–5, 89, 98–101
pullback of, 36, 46
quotient, 41, 55, 111
reducible, 41–8, 54, 91, 96–9, 109–11
restriction of, 36, 43, 47
trivial, 38, 50
Verma, 111
multiplicity
of a weight, 51, 57, 63–4, 76, 79–81, 83–5, 98, 104–5, 107, 110–12
of an irreducible summand, 57–8, 101

normalizer, 32

perpendicular subspace, 60, 67
pure tensor, 4, 62, 65

quotient, 27–9
maximal abelian, 29, 37, 47

rank, 108
representation, 33–40
adjoint, 36–7, 39, 44, 47, 50
equivalence of, 38–40
faithful, 34–7
natural, 35–6
of a Lie group, 1
of GL_n, 1–11, 101
of $\mathfrak{gl}_2$, 8, 19
of $\mathfrak{gl}_n$, 10–11
of S_k, 112–13
representing matrices, 34–40, 46–7
root, 84, 107–9
positive, 108
simple, 84, 107–9
root system, 108

Schur polynomial, 112
Schur's lemma, 68–9
Schur–Weyl duality, 113
semi-direct product, 48, 76
semistandard Young tableau, 112–14
skew-symmetric matrix, 12, 15
$\mathfrak{sl}_2$-triple, 57, 79, 106–9

standard basis
  of $\mathfrak{gl}_n$, 16, 74
  of $\mathfrak{sl}_2$, 18
  of $\mathfrak{so}_3$, 19
strictly upper-triangular matrix, 22, 26, 29, 78, 107
string, 54, 56, 58–9, 61, 63, 66, 76, 87, 94, 113
subalgebra, 11, 21–5, 28–9, 31–2
  Cartan, 108
  complementary, 32
  generated by a subset, 23–4, 31
  of $\mathfrak{gl}_2$, 16, 19, 23–4, 31
  of $\mathfrak{gl}_n$, 21, 24
  of Mat_n, 22
submodule, 41–9
  complementary, 45–6
  generated by a subset, 44, 47
  nontrivial, 41
  trivial, 41
symmetric tensor, 5–7, 64, 74, 93–5, 97, 103–5

tangent space, 11
tensor power, 6, 64, 112
tensor product, 4–7, 61–7, 93–7, 101–5
trace, 12, 14, 38, 74
trace form, 70–2, 74–6
transpose, 2, 4, 14, 61, 67, 79

universal enveloping algebra, 114
  quantized, 113
upper-triangular matrix, 22, 26, 29, 43

Vandermonde determinant, 111

weight
  dominant, 87, 89, 109–11
  for $\mathfrak{gl}_n$, 79–89, 93–8
  for $\mathfrak{sl}_2$, 51, 57, 63
  for $\mathfrak{sl}_n$, 82–5
  fundamental, 82, 89, 110
  higher, 86
  highest, 88, 92, 96–9, 109–11
  integral, 79–83, 87, 109–11
weight lattice, 79
weight space, *see* weight vector
weight vector
  for a simple Lie algebra, 107–8, 110, 113
  for $\mathfrak{gl}_n$, 79–89, 93–5, 98, 100, 102, 112
  for $\mathfrak{sl}_2$, 51–7, 63
  for $\mathfrak{sl}_n$, 82–5
Weyl character formula, 111
Weyl's theorem, 47, 110
  for $\mathfrak{sl}_2$, 55
  for $\mathfrak{sl}_n$, 100

Printed in the United States
by Baker & Taylor Publisher Services